SPECIAL PAPERS IN PALAEONTOLOGY NO. 87

TABULATE CORALS FROM THE GIVETIAN AND FRASNIAN OF THE SOUTHERN REGION OF THE HOLY CROSS MOUNTAINS (POLAND)

BY

MIKOŁAJ K. ZAPALSKI

with 30 figures and 35 tables

THE PALAEONTOLOGICAL ASSOCIATION
LONDON

April 2012

CONTENTS

Page

ABSTRACT 5
INTRODUCTION 5
GEOLOGICAL BACKGROUND 6
MATERIAL, METHODS AND TERMINOLOGY 8
THE SYSTEMATIC POSITION OF TABULATE CORALS 10
MICROSTRUCTURE OF TABULATE SKELETON 10
SPECIES IN TABULATE CORALS 11
SYSTEMATIC PALAEONTOLOGY 12
    Order Favositida 12
        Family Favositidae 12
            Genus *Pachyfavosites* Sokolov, 1952 12
        Family Pachyporidae 13
            Genus *Striatopora* Hall, 1851 13
            Genus *Thamnopora* Steininger, 1831 16
        Family Alveolitidae 23
            Genus *Alveolites* Lamarck, 1801 24
            Genus *Crassialveolites* Sokolov, 1955 42
            Genus *Alveolitella* Sokolov, 1952 52
        Family Caliaporidae 56
            Genus *Caliapora* Schlüter, 1889 56
            Genus *Natalophyllum* Radugin, 1938 58
            Genus *Scoliopora* Lang, Smith and Thomas, 1940 60
        Family Coenitidae 61
            Genus *Coenites* Eichwald, 1829 62
            Genus *Platyaxum* Davis, 1887 63
            Genus *Roseoporella* Spriesterbach, 1935 64
    Order Syringoporida 66
        Family Syringoporidae 66
            Genus *Maksymilianites* Zapalski and Nowiński, 2005 69
            Genus *Sapounofouskilites* gen. nov. 70
            Genus *Syringopora* Goldfuss, 1826 72
        Family Multithecoporidae 74
            Genus *Syringoporella* Kettner, 1934 74
    Order Auloporida 75
        Family Auloporidae 75
            Genus *Aulopora* Goldfuss, 1826 76
        Family Aulocystidae 79
            Genus *Adetopora* Sokolov, 1955 79
INTRACOLONIAL VARIATION 80
PARASITISM IN TABULATE CORALS 82
STRATIGRAPHIC AND PALAEOGEOGRAPHIC DISTRIBUTION OF TABULATES IN THE KIELCE REGION 86
CONCLUSIONS 90
ACKNOWLEDGEMENTS 91
REFERENCES 91

[Special Papers in Palaeontology, 87, 2012, pp. 5–100]

Abstract: Givetian and Frasnian tabulate corals from the southern region of the Holy Cross Mountains, Poland, are described. Both Givetian and Frasnian tabulate faunas from the study region are dominated by alveolitids and comprise 52 species (Favositida: 40 species, Syringoporida: 6 species, Auloporida: 6 species). A new genus belonging to Syringoporida is proposed – *Sapounofouskilites* gen. nov. – and five new species are erected (Favositida: *Striatopora sciuricauda* sp. nov., *Alveolites? obtortiformis* sp. nov., *Crassialveolites oliveri* sp. nov., *Roseoporella heuvelmansi* sp. nov.; and Auloporida: *Aulopora slosarskii* sp. nov.). Study of the intracolonial variation in tabulates shows that minimal and maximal corallite lumen diameters and pore diameters are the best taxonomical discriminators for Alveolitidae and Coenitidae, while the double wall thickness and tabulae spacing are less useful characters. Moreover, alveolitids and coenitids show overall greater intracolonial variation than, for example, heliolitids. The tabulate endobiont *Chaetosalpinx? plusquelleci* isp. nov. is newly described. Study of tabulate endobionts – that is, *Chaetosalpinx? plusquelleci*, *Helicosalpinx* cf. *asturiana* Oekentorp and *H.* isp. – show that these were parasites of tabulate corals. Givetian and Frasnian tabulate faunas from the Kielce (southern) Region of the Holy Cross Mountains are dominated by Alveolitidae.

Key words: Tabulate corals, Devonian, intracolonial variability, palaeoecology, species definition.

THE Devonian deposits of the Holy Cross Mountains, Central Poland (Fig. 1), have yielded abundant fossils belonging to most of the Palaeozoic marine animal groups. These fossils have been studied since the mid-nineteenth century, beginning with Pusch (1837) and followed by Gürich (1896) and Sobolev (1904); major contributions followed in the mid-twentieth century thanks to several field campaigns, the best-known of which is that at the Grzegorzowice-Skały section about 1947 (e.g. Biernat 1953, 1959, 1964, 1966; Kielan 1954; Pajchlowa 1957; Stasińska 1958; Kiepura 1965, 1973).

The bioconstructors (stromatoporoids, tabulate and rugose corals) from the Kielce Region – the southern part of the Holy Cross Mountains – are important rock-forming fossils. Tabulate corals are particularly common; nonetheless, research undertaken on this group in the past was very fragmentary. Roemer (1866) followed by Gürich (1896) recognized these organisms in the Holy Cross Mountains for the first time and recorded several species. The first paper devoted solely to tabulates from the Holy Cross Mountains was that by Stasińska (1953) on representatives of the genus *Alveolites*. This paper deals with the fauna from the Kielce Region and describe and illustrate several species, most of them known previously from the Franco-Belgian Ardennes (Lecompte 1933, 1939; Stasińska 1953). Later, Stasińska (1954, 1958, 1969a, 1974) dealt mainly with fauna from the Łysogóry Region in the northern part of the Holy Cross Mountains. Sarnecka (1987, 1997) and Zapalski (2003, 2005a, b) completed studies of the tabulate faunas from the Łysogóry Region; Zapalski (2007a, 2009) studied the parasites of these corals, and their growth patterns have been documented by Zapalski *et al.* (2012). Nowiński (1970) published a paper on a new genus of syringoporid (Zapalski and Nowiński 2005) from Sowie Górki (Kielce Region) and later (Nowiński 1992) a preliminary study with a list of species from the Holy Cross Mountains (focusing principally on faunas from the Kielce Region). In addition, several new taxa were described.

The tabulate corals from the Kielce Region have been neglected in comparison with other large groups of Devonian benthic animals from the Kielce Region, such as stromatoporoids (Kaźmierczak 1971) and rugose corals (Wrzołek 1992). Hence, the present study aims to fill that gap through the following objectives:

1. presentation of taxonomic monograph of tabulates (excluding heliolitids) from the Kielce (Southern) Region;
2. description of intracolonial variability in selected taxa. This study was conducted with special regard to members of the family Alveolitidae, one of the most common Devonian tabulate families; and
3. identification of interactions with endobionts of tabulate corals.

As stated above, the first large topic under study is a description of taxonomic composition of tabulate faunas from the Devonian of the Kielce Region. In addition, it was necessary to review several species described by Lecompte (1933, 1939) from the classical sites of Ardennes, because the definition of the species often is inadequate; the identification of Polish material was impossible without revision of the type material. Five species of 52 described here are new; also, a new genus is erected. Special attention was paid to the intracolonial variation and, in particular, to the evaluation of commonly used biometrical parameters as taxonomic discriminators. This type of study is performed here on alveolitids and coeni-

by MIKOŁAJ K. ZAPALSKI

Faculty of Geology, University of Warsaw, Żwirki i Wigury 93, 02-089 Warszawa, Poland; e-mail: m.zapalski@uw.edu.pl

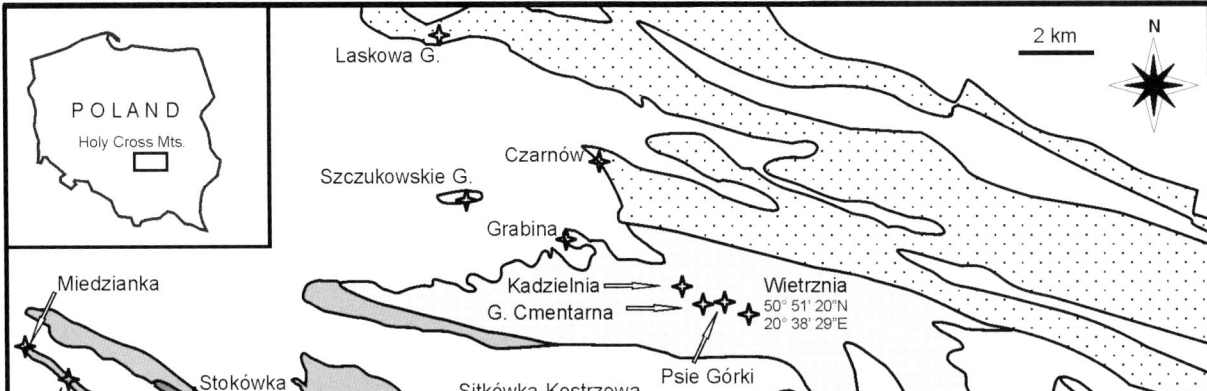

FIG. 1. Geological map of the western part of the Holy Cross Mountains showing the location of the fossiliferous sites in a palaeogeographic setting of the Frasnian basin (after Racki 1992*a*). Trzemoszna is out of the range of this map. Bol., Bolechowice.

tids. Endobionts hosted by *Alveolitella* and *Pachyfavosites* are described here for the first time; moreover, *Helicosalpinx* organisms are recorded from Poland for the first time. Finally, remarks on patterns of stratigraphic distribution of tabulate faunas in the Kielce Region are given. Recently, Zapalski *et al.* (2012) published a complementary study of alveolitid growth patterns based on material described in this monograph.

GEOLOGICAL BACKGROUND

The Holy Cross Mountains (= Góry Świętokrzyskie in Polish) are a range of hills in Central Poland (Fig. 1). They primarily consist of a Palaeozoic core surrounded by Permo-Mesozoic and Miocene deposits and can be further subdivided into two units: Kielce Region (southern) and Łysogóry Region (northern). The latter subdivisions were first recognized by Michalski (1888*a*, *b*) and defined by Czarnocki (1950); they have both tectonic (border: Holy Cross Fault) and palaeogeographic (border: south from the Holy Cross Dislocation) implications (Stupnicka 1992; Mizerski 1995; Szulczewski 1995; Lamarche *et al.* 1999). The two regions are regarded as terranes, and Nawrocki and Poprawa (2006) outlined their early history. Their conjunction probably took place before the Emsian, and Lewandowski (1993) suggested the Silurian, at which date broadly distributed clastic deposits became unified in the entire area.

The Devonian in the Holy Cross Mountains

Devonian strata are exposed in both the Łysogóry and Kielce Regions of the Holy Cross Mountains, and Szulczewski (1995) described the Devonian succession. The basin of Northern Region (Łysogóry Region) was in the Devonian deeper than Southern (Kielce) Region, and therefore, Racki (1992*a*) distinguished Łysogóry palaeolow and Kielce palaeohigh; between these palaeobathymetric zones, also Kostomłoty Transitional Zone can be distinguished. Chęciny-Zbrza basin, south from Kielce Region, is sometimes included in the latter, sometimes regarded as separate, southernmost basin.

Kielce Region. Lower Devonian strata in the Kielce Region overlie discordantly on Cambrian and Silurian deposits (Mizerski 1995, 2004; Szulczewski 1995) and consist of clastic sediments, namely conglomerate at the base, followed by fine-grained clastics (Szulczewski 1995). The Emsian deposits in the Southern Region represent shallow environments (presumably lagoonal), and they are rich in vertebrate remains (Tarlo 1964, 1965), whereas deposits of the Northern Region yielded open marine biota (e.g. Łobanowski 1971). These shallow water lagoonal to marine deposits are overlain by the Middle Devonian eogenetic dolostones of the Wojciechowice Beds in the Northern and Southern Regions.

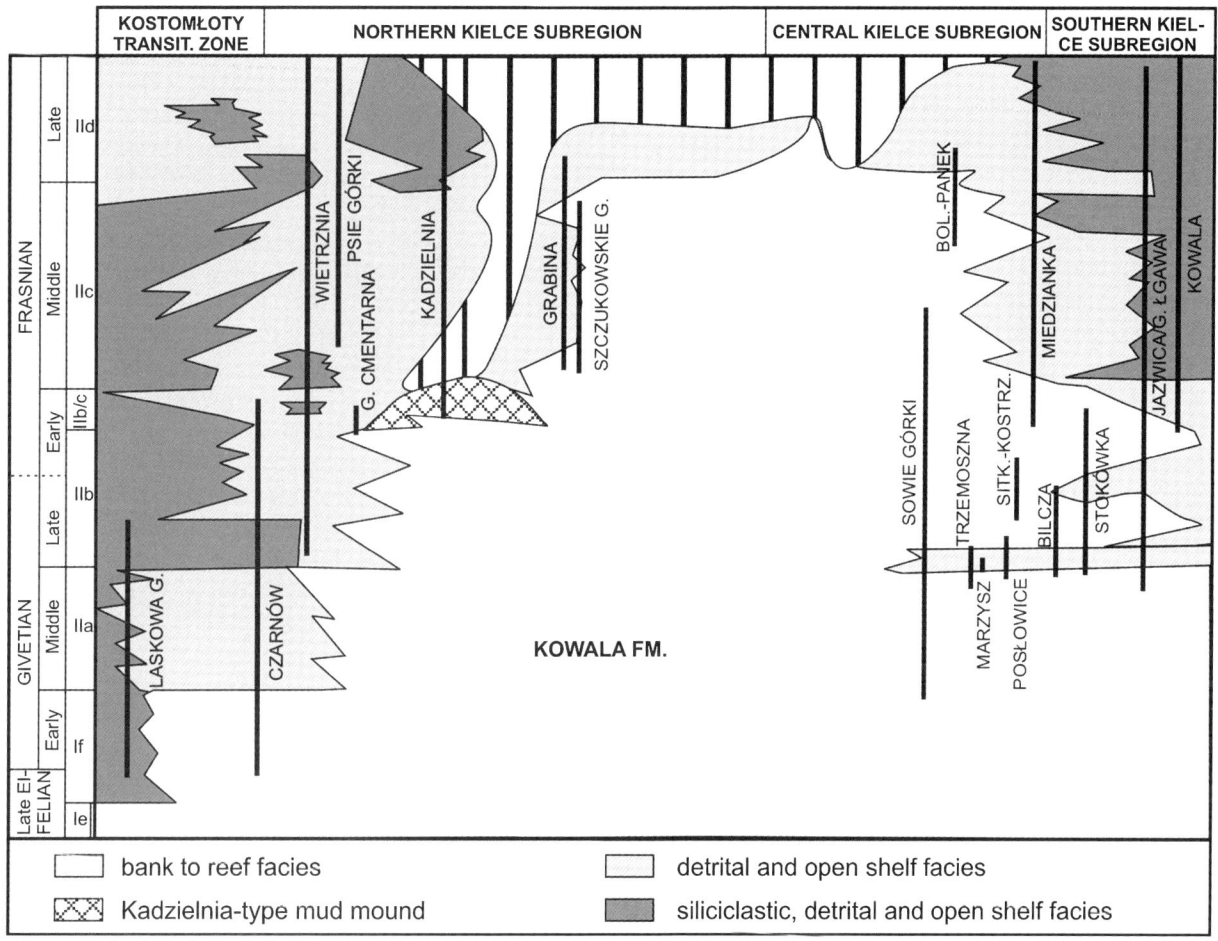

FIG. 2. Facies distribution and sections that provided the samples for the present study (after Racki 1992*a*).

Wojciechowice Beds in Southern Region are overlain by limestone referred to as the Kowala Formation (Narkiewicz *et al.* 1990). The base of the limestone is correlated with the Ie transgressive-regressive cycle of Johnson *et al.* (1985; Szulczewski 1995). Carbonate platform development is divided into three successive steps: (1) the first phase (late Eifelian – early Givetian) is characterized by low depositional topography, with accumulation restricted to shallow (tidal and subtidal) environments and a nearly absence of buildups, (2) the second phase (Givetian) – the carbonate bank phase – when biostromal stromatoporoid-coral limestone accumulated; the drowning of the stromatoporoid bank is correlated with the IIb T-R cycle *sensu* Johnson *et al.* (1985; see Racki 1997; Pisarzowska *et al.* 2006). The limestone deposition continued until the Frasnian, when (3) bioherms were formed, making up the third phase (Dyminy reef complex of Racki 1992*a*). These shallow water sediments are extremely rich and with diverse faunas, especially stromatoporoids (Kaźmierczak 1971), sponges (Hurcewicz 1992; Rigby *et al.* 2001), tabulate (Stasińska 1954, 1958; Nowiński 1970, 1992) and rugose corals (e.g. Różkowska

1953, 1979; Wrzołek 1992), brachiopods (Racki 1992*b*), rare trilobites (Chlupáč 1992), ostracodes (Olempska 1979; Malec and Racki 1992), tentaculites (Hajłasz 1992) and gastropods (Krawczyński 2002, 2006). At the end of the Frasnian, the carbonate platform drowned during the IId transgressive pulse *sensu* Johnson *et al.* (1985), which correlates with the *Kellwasser* events (Racki 1997) and deeper marine deposits dominated. However, in some places, regional tectonic movements caused subaerial exposure of the succession (e.g. Ostrówka in the western part of the Southern Region; Szulczewski *et al.* 1996). Famennian deposits consist mainly of shale and limestone, containing rich ammonoid, conodont, brachiopod and trilobite faunas (Czarnocki 1989; Berkowski 1991; Dzik 2002, 2006, Halamski and Baliński 2009), with some rugose corals and a single known specimen of tabulate coral *Yavorskia* sp. (Berkowski 2002, p. 8; Zapalski and Berkowski in press). Pisarzowska *et al.* (2006) presented a detailed outline of the development of the carbonate platform.

Racki (1992*a*) divided the Kielce Region into four subregions (Fig. 1): Northern, Central, Southern and Chęciny-Zbrza. These subregions refer to different bathymetric zones of the Devonian basin. The Central Subregion was the shallowest one, while the others were deeper, with the Chęciny-Zbrza Subregion being the deepest. The division into subregions reflects a symmetric Frasnian reef, named the 'Dyminy reef' by (Racki 1992a), with the Northern and Southern subregions representing the foreslope. The material analysed in this study comes from the first three subregions and from the Kostomłoty Transitional Zone, a transitional zone between the Łysogóry and Kielce Regions. The schematic facies distribution in the Kielce Region is shown in the Figure 2.

MATERIAL, METHODS AND TERMINOLOGY

Collection sites

In total, 19 localities provided material for this study. The site descriptions, stratigraphic and lithologic details are presented by Racki (1992a). Racki (1992a) divided the carbonate platform succession and surrounding rocks into informal lithological sets that are used in local stratigraphy. The lithological set (A, B, C, etc.) given in the species descriptions follows the terminology of Racki (1992a, pp. 171–180). The geographical location of these sites is shown on Figure 1, while the stratigraphic positions are given in the Figure 2.

Material

The material has been collected by A. Nowiński, A. Stasińska and the author. At certain localities, the samples were collected bed-by-bed with several tens of samples from each bed. At other localities, the bed-by-bed collecting was not possible, and the specimens were collected from the rubble. This procedure was caused by (1) technical difficulties in the field extracting corals from massive and very hard limestones, for example in the Kadzielnia Quarry, but also (2) collecting *in situ* samples is banned by law at some quarries, for example the Laskowa Quarry, which hampered bed-by-bed sampling. The origin of the loose samples in the succession, however, could often be identified in the quarries, but in the cases of doubt on their origin (i.e. bed), a question mark has been added to the sample.

Methods

The coralla of tabulate corals were investigated in this study mainly by (1) thin sections, (2) serial acetate peels (grinding) and (3) ultrathin sections. Over 600 thin sec-

tions, 20 ultrathin sections and about 100 serial acetate peels were prepared for this study, in addition about 700 thin sections from A. Nowiński's and about 300 thin sections from A. Stasińska's collections were used. Nowiński's and Stasińska's sections have usually a surface of 25×25 mm, while those prepared by the author are up to 100×120 mm (for alveolitids). In total, about 1100 thin sections were suitable for study, whereas others were not useful, owing to obliquities in cutting or poor preservation.

Whenever it was possible, the thin sections were oriented, that is, transverse sections perpendicularly to the axes of corallite growth and longitudinal sections, which are concordant to the axes of corallite growth. In branching corals, also tangential sections were made. They have been positioned as close as possible to the external surface.

Alveolitids are tabulates with corallite bilateral symmetry. The ideal situation is where four sections can be prepared (Hladil 1981a): perpendicular (transverse) and three longitudinal (horizontal, vertical and oblique). While in selected, small specimens this can sometimes be done, in most of colonies this is technically impossible because corallites are meandering and changing their curve every few millimetres. The descriptions of anatomical features are inferred from normal transverse and longitudinal sections.

The serial acetate peels were prepared in order to follow the cyclomorphic changes, and they allowed recognizing intracolonial variations. The ultrathin sections are prepared at first by progressive grinding on normal abrasive materials, then on aluminium suspension liquid and finally with grains of 200 Å in diameter.

Measurements and abbreviations used

As there is no standard procedure for taking measurements for biometrical study in tabulate corals, all measurements are defined below (Fig. 3):

Corallum size. The measurements are with an accuracy of 1.0 mm.

Branch diameter. In branching tabulates, the measurements are with an accuracy of 1.0 mm.

Corallite diameter (CD). In corallites polygonal in cross section, usually two measurements were taken: (1) between the most distant angles and (2) perpendicularly to the axes of this measurement. When median line was present, the measurements were taken between facing median lines (Fig. 3C); otherwise, the lumen (= visceral chamber) diameter (LD) was measured in an analogous way (Fig. 3A). Also in alveolitid corallites (that is elongated in cross

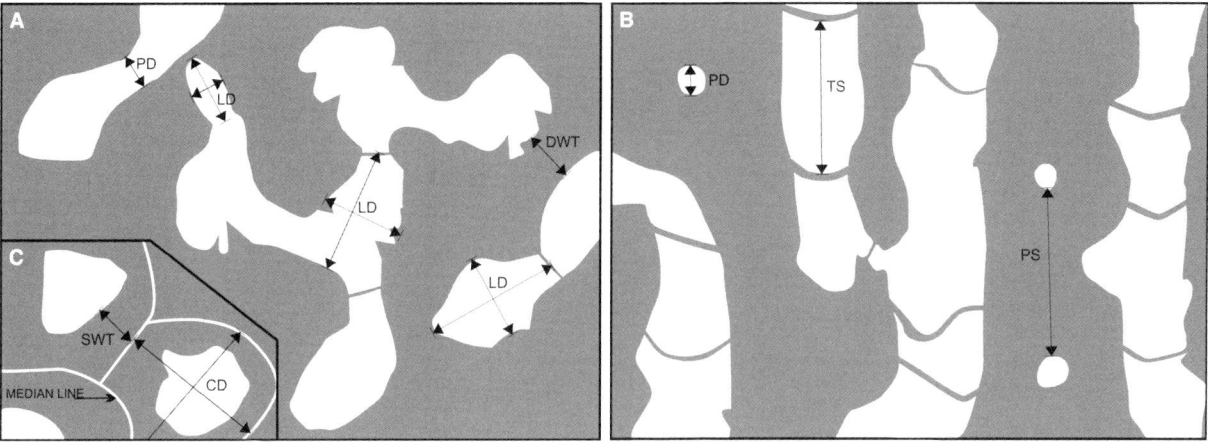

FIG. 3. A scheme of measurements in alveolitid corals. A–B, Transverse section (A shows a case with median line invisible, B the case when median line permits to delimit individuals), C, longitudinal section. CD, corallite diameter; DWT, double wall thickness; LD, lumen diameter (Max LD, Min LD: maximal and minimal lumina diameters, respectively); PD, pore diameter; PS, pore spacing; SWT, single wall thickness; TS, tabulae spacing.

section), two measurements were taken: (1) the longest and (2) the shortest *lumen diameters* (= Max LD and Min LD, respectively).

Corallite wall thickness (SWT) was measured when the median line was visible (Fig. 3C); in other cases, the *double wall thickness* (DWT) was measured (Fig. 3A).

Pore diameter. Pore diameter (PD) was measured on tangential longitudinal sections of the wall (Fig. 3B); if such a section could not be obtained, their diameter was measured on cross section (as indicated as 'PD' on Fig. 3A). It must be however stressed that in such cases, the resulting value is not the true diameter.

Pore spacing. Pore spacing (PS) was measured on the tangential sections of the wall. The spacing was measured from the edge to the edge of the pore (Fig. 3C).

Tabulae spacing. Tabulate spacing (TS) was measured according to the scheme of Zapalski *et al.* (2007*a*) that is on longitudinal section in the central zone of corallite. All measurements if otherwise not indicated are with an accuracy of 0.02 mm.

Terminology used in the systematic descriptions

The terms used to describe corallites are geometric (Hill 1981). The term 'shield-like corallite' is used to describe corallites, where the upper wall is divided longitudinally into two parts, concave to the corallite lumen, and the lower wall is strongly convex to the corallite. For the corallum structure, three terms are used (after Hill 1981, modified):

Cerioid – where corallites are in connection throughout their length.
Fasciculate – where corallites are not in touch, but connected by platforms, tubuli and other connecting elements.
Reptant – the corallites are connected only by budding pore; thus, the only connection is ancestor-descendant, with no vertical connections between individuals of the same generation.

For the corallum morphology, the following terms are used:

Laminar – the horizontal extension of the colony is significantly larger than vertical.
Discoidal – the horizontal extension of the colony is larger than vertical.
Columnar – the horizontal extension of the colony is significantly smaller than vertical.
Bulbous – the base of the corallum is smaller then upper part and general shape is bulb-like.
Branching – bushy coralla with branches like in tree.
Palmate – flat, lamellar coralla with anastomosing branches; also, the term *cambered* may be used if they are irregularly folded. All these forms may have protuberances on the colony surface; in such a situation, adjective *aberrant* is used. In all cases where the exact form of corallum is unknown (e.g. deduced from thin section), the term *massive* is used; laminar coralla can be recognized even in thin sections. One must keep in mind that these terms are descriptive and all possible transitional forms are present.

The intracolonial variation. In selected cases, when material allowed (good preservation, good quality and correct orientation of thin sections), a survey on intracolonial variation was undertaken. Tables showing variations of different characters within coralla (one table per corallum) were prepared, with indicated minimal/maximal values, means, standard deviations and number of taken measurements. In each possible case, at least 30 measurements of each character were taken. The values specified in species descriptions provide therefore idea about intraspecific and intracolonial variations (minimal and maximal values were given, as well as values of mean and standard deviation). Moreover, a 'coefficient of variation' (Simpson *et al.* 1960; Young and Elias 1995; Mõtus 2006) was counted for each corallum and each character. Coefficient of variation (*V*) was calculated as follows (after Simpson *et al.* 1960, modified):

$$V = \text{standard deviation/mean}.$$

This coefficient allows comparisons between characters with very different values. Such studies were performed mainly on alveolitids and coenitids.

Repository. The specimens analysed in the present study are housed at the Institute of Palaeobiology, Polish Academy of Sciences (Warsaw), under the repository number ZPAL T.25. All inventory numbers (if not stated otherwise) refer to this collection. Specimens from other collections housed at the Institute were also taken in account (abbreviation ZPAL). Specimens from the collections of Musée de l'Institut royal des Sciences naturelles (Bruxelles) appear under the numbers starting from MRHN. Numbers of samples (and/or thin sections) may appear in descriptions of more than one species, as (especially for large thin sections) often more than one corallum is cut at once.

THE SYSTEMATIC POSITION OF TABULATE CORALS

The systematic position of tabulate corals has been discussed for a long time. Milne-Edwards and Haime (1850) placed this group within corals, but Kirkpatrick (1911, 1912*a, b*) questioned this idea.

Some authors (Flügel 1976; Stel 1978; Kaźmierczak 1984, 1989, 1993, 1994) assigned tabulate corals (mainly favositids, but also heliolitids) to sclerosponges owing to the presence of 'spicules' within the corallite wall. Later, Oekentorp (1985), Reitner (1989) and Wood *et al.* (1990) demonstrated that the 'spicules' are in fact remains of postmortem borings or simply diagenetic artefacts (Scrutton 1997).

In contrast, Hartman and Goreau (1975) concluded that favositids were corals. Oliver (1979) argued that tabulates are corals, because favositids have clearly delimited individuals (see also Oliver 1986). Several authors placed selected tabulatans within bryozoans, that is, auloporids (Fenton and Fenton 1937); or coenitids (Brood 1970).

The cnidarian affinity of tabulates was emphasized by Copper (1985) who described fossilized polyps of *Astroce-*

rium? sp. (originally attributed by Copper to *Favosites*; see Oekentorp and Stel 1985; material from the Silurian of Anticosti Island, Canada), and his discovery gave strong support for coral affinities of tabulates. Recently, Chatterton *et al.* (2008) and Dixon (2010) showed that sclerites were present in the Silurian tabulate material from Anticosti Island. Similar sclerites occur in modern organ-pipe corals (Octocorallia; Chatterton *et al.* 2008), and this discovery suggests relationship between these groups. Moreover, the arrangement of the spicules emphasizes the dodecal symmetry characteristic of Tabulata. Dodecal symmetry occurs also in the microstructure of auloporids and syringoporids and provides strong argument in favour of anthozoan affinities of tabulate corals (Mistiaen 1989). Scrutton (1987, p. 490) gave the most comprehensive discussion on the systematic position of favositids and concluded the following:

'There appears to be no single item of evidence in favour of the favositids being sponges, other than a very gross morphological similarity with the tabulosponges. Mural pores and septal structures both show features more strongly related to other tabulates and to the Rugosa respectively (…). To maintain any claim for sponge affinities for these extinct organisms, not only must Copper's polyps be explained away, but some new and convincing positive evidence must be forthcoming'.

Subsequently, Scrutton (1997) emphasized the discovery of fossilized polyps (Copper 1985) as an argument for their biological affinities. The present author follows the arguments of Scrutton (1987, 1997) and includes tabulates within the corals.

MICROSTRUCTURE OF THE TABULATE SKELETON

The microstructures of tabulate and other corals have been observed for a very long time. Until 1970, the microstructures of tabulate skeleton had been studied by using traditional (petrographic) thin sections, but Lafuste (1970, 1980) invented ultrathin sections, a method allowing preparing sections of about 2 μm thick, which observed in polarized light well show borders of crystals.

The taxonomical use of microstructures, observed both in ultrathin sections and in standard-thin sections, have been discussed by numerous researchers. Hill and Butler (1936) were the first to observe microstructures that they recognized as diagenetically changed. Later, other authors (e.g. Kato 1963; Lafuste 1979*b*) made more observations, sometimes assuming a primary and sometimes a secondary (diagenetic) origin of observed microstructures. As stated above, they were regarded by a group of researchers as primary microstructures (e.g. Lafuste 1958*a*, 1959,

1963, 1978; Lafuste and Debrenne 1970; Plusquellec and Tchudinova 1977; Plusquellec and Sando 1987; Lafuste and Tourneur 1991), therefore regarded as biogenic. Lafuste (1958*a*, *b*, 1959, 1978, 1979*a*) described various types of microcrystals: lamellae, microlamellae, pseudolamellae, desmids and others. On the basis of these structures, far-reaching conclusions were made, mainly concerning the taxonomy and systematics (e.g. Lafuste 1958*b*, 1983, 1984, 1986; Lafuste and Plusquellec 1985; Plusquellec *et al.* 1997, Plusquellec 2007), not only within the group of tabulates, but hydrozoans as well (Lafuste 1979*a*). The other group of researchers came to different conclusions: they assumed that the majority of these microstructures were secondary (Kato 1963; Oekentorp 1972, 1980, 1989, 2001; May 1993*a*; Sorauf 1997; Oekentorp and Brühl 1999; Sorauf and Webb 2003), in other words diagenetic.

May (2006) has recently triggered a new discussion on the validity of microstructures in taxonomical investigations and their origin – whether they are primary or secondary (see May 2007; Oekentorp 2007; Plusquellec and Fernández-Martínez 2007). Oekentorp (2007) advocated the 'secondary interpretation' of microstructures, and his paper contains two important statements: 'Microstructures occurring in Palaeozoic corals which do not correspond to fibrous or trabecular structures can easily be explained as diagenetic changes on the basis of crystallographic regularities' (Oekentorp 2007, p. 95) and 'publications with controversial interpretations have been neither critically argued nor cited. Instead, these secondary microstructures were used directly for systematic and phylogenetic purposes' (Oekentorp 2007, p. 96). Moreover, in favour of diagenetic alteration of original microstructure can be an argument of Perrin and Cuif (2001) who discovered that the earliest changes in the microstructure of the skeleton of modern corals begin as early as few years after formation of the hard tissue (see also Perrin and Smith 2007).

Presented points of view seem to be extremely different, but it appears that both may be partly right. As shown by Perrin and Cuif (2001), the truly biogenic microstructure is often altered, but the resulting structures are always in strong connection with the original ones – therefore, the most important outlines of the primary microstructure are retained. To support such a statement, results of Nothdurft and Webb (2007) can be reminded, where 'shingles' were obscured by diagenesis, but they were preserved as recognizable units. Cretaceous *Stephanocyathus* can also be brought to mind, where trabeculae are preserved despite diagenesis where the original aragonitic skeleton was altered (Stolarski 1990). Also, *Multithecopora* described by Nowiński (1991) displays two-layered stereoplasm with distinct microstructures (similarly as *Sapounofouskilites minimus* described below); the primary microstructure was most probably

changed during the diagenesis, but a bilayered original structure of the corallite wall is reflected here.

Because of the poor preservation of the investigated material in the present work, skeletal microstructures are generally omitted unless the preservation of material allows such studies.

SPECIES IN TABULATE CORALS

The species concept of Mayr (1942) is supposed by a considerable number of biologists to be classic, yet in fact it is based on a limited empiric material. It cannot be directly applied in palaeontology for obvious reasons; what is more, it should be treated rather as a theoretical concept than a ready-to-apply working definition (Hey 2006). In practice, palaeontologists (and most of biologists) apply the morphological concept of species (e.g. Budd 1993; Foote and Miller 2007). In the morphospecies approach, biometry plays an important role (Imbrie 1956). This has also the most often been the case in tabulate corals, the most commonly studied biometrical parameters being: corallite size, thickness of intercorallite wall, size of connecting elements (e.g. pore diameters), and spacing of connecting elements and tabulae (e.g. Sorauf and Stein 1993; Young and Elias 1993; Bae *et al.* 2006). It needs to be emphasized that some of these elements remain under strong environmental influence (e.g. Watkins 2000), and their usefulness for taxonomic purposes remains questionable (Young and Elias 1993, see also below in chapter on intracolonial variation). On the species-level taxonomy, besides biometric (quantitative) characters also qualitative features are important, like shape of corallites and connecting pores, completeness of tabulae, or corallum form (Sorauf and Stein 1993), as many taxonomically important features cannot be represented numerically (Imbrie 1956). Following the above-mentioned author (Imbrie 1956, p. 221), it needs to be strongly accentuated that 'taxonomic discrimination is properly based on a combination of morphologic, geographic, and stratigraphic evidence. Evaluation of this sort is thus fundamentally nonstatistical and ideally should never rest on biometrical data alone'.

In tabulate corals, it has been generally agreed that if a single given biometrical character (e.g. corallite size) or its derivate (e.g. corallite diameter to wall thickness ratio) forms an isolated cluster, then the analysed species is distinct from the other representatives of a given genus (see case studies by Bae *et al.* 2006, 2008; Mõtus 2006).

The concept of morphospecies in fossil corals can be criticized from a neontological perspective. Recent corals show remarkable environment-dependent plasticity (Budd

1993), especially in terms of colony morphology (Hoeksema 1993; Veron 1993, 2007; Todd 2008). Genetic intraspecific variability is also an important factor causing morphological differences in scleractinian species (Todd 2008). Although the role of molecular data has significantly increased recently (e.g. Fukami *et al.* 2004), micromorphological characters (like shapes and distributions of septal teeth and granules) can still be useful in taxonomy (Budd and Stolarski 2009). Systematics of scleractinians require therefore multiple criteria – morphological, molecular and reproductive (Lang 1984; Zlatarski 2007; an example of a case study is given by Maté 2003); in certain cases, the skeletal and molecular taxonomies seem to be consistent with each other (Potts *et al.* 1993), and in certain others, they do not seem so (Miller and Benzie 1997). Nonetheless, most of characters used in modern coral taxonomy cannot be applied in palaeontology. In other groups of colonial animals, morphospecies seem to be in general genetically distinct (e.g. bryozoans: Jackson and Cheetham 1990).

A species is often considered as the sole taxonomical unit existing in nature; moreover, it is generally thought that species are true biological units (e.g. Veron 1993, p. 61); some authors, however, consider only syngameons (units composed of genetically related organisms that may or may not be morphologically similar) as biological units (Veron 2007, p. 511); yet syngameons may (according to an interpretation) belong to several species (González-Forero 2009). Such a puzzle can lead to the definition expressed by Kitcher (1984, p. 308): 'Species are those groups of organisms which are recognized as species by competent taxonomists'.

It is difficult to define what a species in tabulate corals is. It is of course a morphospecies, usually (but not always) defined on biometrical basis. Coral researchers are in this lucky situation that even a single colony provides a population of individuals sharing the same genotype, and consequently, a biometrical study can be performed on a single specimen. One colony, however, does not give information on intraspecific (extracolonial) variability.

A good example of a species defined on biometrical basis is *Striatopora sciuricauda* sp. nov. (see below). In this species, most of the parameters are overlapping with closely related species, but there is always one parameter separating this species from others. *Alveolites suborbicularis* Lamarck and *Alveolites compressus* (Milne-Edwards et Haime) are species with very similar biometrical parameters, but the shape of corallites (a qualitative character) slightly differs one from the other.

There will never be a total agreement between 'competent taxonomists' whether a given specimen belongs to one species or another; the concept of Mayr (1942) is per-

haps correct, but the one of Kitcher (1984) is easier to use in practice. Scientists aim to obtain objective data, but in a taxonomist's work, besides the 'hard data' intuition is (unfortunately) still very important; what is more, some nonbiometrical characters by some may be considered as taxonomically important, by some others as unimportant.

SYSTEMATIC PALAEONTOLOLOGY

Remarks. The new taxa described by Nowiński (1970, 1992) are neither redescribed nor re-illustrated in the present paper, unless new material or new data were added or better illustrations are provided.

Class ANTHOZOA Ehrenberg, 1834

Subclass TABULATA Milne-Edwards and Haime, 1850

Discussion. The classification presented below follows the classification of Hill (1981), with the exception of Auloporida/Syringoporida. Order Auloporida Sokolov, 1950 (*sensu* Hill 1981) is divided here into two separate orders: Auloporida Sokolov, 1950 (*sensu* Sokolov 1950, 1962*a, b*) and Syringoporida Sokolov, 1947 (*sensu* Tchudinova 1986). In the present author's opinion, the main feature differencing these two orders is the development of the coralla: auloporids have two dimensional skeletons with lacking connecting elements, while the syringoporids have the coralla developed in three dimensions with numerous connecting elements, such as tubuli and platforms (see Sokolov 1962*a, b*; Hill 1981).

Order FAVOSITIDA Wedekind, 1937
Family FAVOSITIDAE Dana, 1846
Subfamily PACHYFAVOSITINAE Mironova, 1965

Genus PACHYFAVOSITES Sokolov, 1952

Type species. Calamopora polymorpha var. *tuberosa* Goldfuss, 1826, Middle Devonian, Eifel Mts.

Pachyfavosites polonicus Nowiński, 1992

*v. 1992 *Pachyfavosites polonicus* sp. n. A. Nowiński, p. 202, fig. 2A–B.
v. 2003 *Pachyfavosites polonicus* Nowiński; Nowiński, p. 130, pl. 68, fig. 3, pl. 69, fig. 2.

Type material. Specimen ZPAL T.18 24/2, Institute of Paleobiology PAS, Warsaw (Nowiński 1992, fig. 2A–B).

Material. Sowie Górki (set G): three coralla, six thin sections (ZPAL T.25 SG 200–205); Posłowice: one corallum, two thin sections (ZPAL T.25 PW 01–02).

Remarks. Nowiński (1992) fully described and illustrated this species. He also mentioned 'commensal worms' occurring within the coralla of *Pachyfavosites polonicus.* These worms belong to *Chaetosalpinx? plusquelleci* isp. nov. and are described in the chapter on parasitism.

Occurrence. Central Kielce Subregion (Middle Givetian–Frasnian).

Superfamily PACHYPORICEAE Gerth, 1921
Family PACHYPORIDAE Gerth, 1921

Genus STRIATOPORA Hall, 1851

Type species. Striatopora flexuosa Hall, 1851 from the Middle Silurian of New York, USA.

Remarks. The genus *Striatopora* is an unusual pachyporid owing to its bimorphism of corallites (Oliver 1966). Small and badly preserved specimens can be easily mistaken with species assigned to the genus *Gracilopora*, from which it differs by the arrangement of corallites in branch. *Striatopora? tenuis* Lecompte and *Striatopora? enigmatica* Nowiński show some intermediate characters between genera *Striatopora* and *Gracilopora* (see discussion on *Thamnopora* below).

Tourneur (1985, p. 55) noticed that very distinct microstructures characterize the two genera *Striatopora* and *Thamnopora* and suggested that they should not be classified within the same family. In the present work, these two genera are classified together as members of Pachyporidae, following Hill (1981).

Striatopora sciuricauda sp. nov.
Figures 4A–F, 5A–B

    vp. *1992 Thamnopora cervicornis* Lecompte; Nowiński, p. 188.
    ? 1992 *Hillaepora circulipora* (Kayser); Nowiński, p. 190, fig. 4 C–G.
    vp. 2003 *Thamnopora cervicornis* Lecompte; Nowiński, p. 135, non pl. 73, figs 1–3.
    ? 2003 *Hillaepora circulipora* (Kayser); Nowiński, p. 133, pl. 79, figs 6–7.

Derivation of the name. From Latin *Sciurus*, that is, the generic name of a squirrel; Latin *cauda* meaning tail, because tangential sections of these corals resemble squirrel tails.

Type strata. Laskowa Beds, Middle Givetian.

Type locality. Laskowa Góra Quarry, Holy Cross Mountains, Poland.

Holotype. Corallum ZPAL T.25 LSG 001.1–4 (Fig. 4A–E); four thin sections.

Additional material. Laskowa (set A, B): three specimens, four thin sections (ZPAL T.25: LSG 002, LG001–002, T.XVIII–12/6); Marzysz: one corallum, six thin sections (ZPAL T.25 M017–021, M024).

Diagnosis. Coralla branching, up to 5–8 mm in diameter. Calyces polygonal, shallow, with lower wall bent, forming a lip. Corallites polygonal in the branch centre, with walls strongly bent in the branch periphery; they reach the corallum surface at acute angle. The corallite diameters 0.40–0.76 × 0.58–0.88 mm in the central parts of branch. Walls even, in the axial part 0.08–0.14 mm thick, up to 0.20 near the surface. Tabulae are few, twisted or strongly inclined, and irregularly spaced. Connecting pores are round, 0.14–0.18 mm in diameter. Septal spines conical, distributed irregularly.

Description. Coralla branching, small, 5–8 mm in diameter, round or slightly oval in cross section. Calyces shallow, subpolygonal, transversally elongated, with lower wall rounded, bent, forming a lip. Corallites long, irregularly polygonal in the branch centre, with walls strongly bent and thickening in the branch periphery, so they become nearly round. Corallites reach the corallum surface at acute angle close to 45 degrees. In the central parts of branch corallite, diameters variable, 0.40–0.76 × 0.58–0.88 mm. Central and peripheral zones cannot be delimitated. Walls even, with moderate thickening towards the corallum surface. Their thickness in central part of the corallum varies 0.08–0.14 mm, most often 0.10 mm, thickening in peripheral part to about 0.20 mm. Median line thin, dark, continuous in axial part of corallum, becoming discontinuous towards the corallum surface. Tabulae not numerous, twisted or strongly inclined, thin, spaced irregularly every 0.32–1.04 mm. Connecting pores uniserial, round, 0.14–0.18 mm in diameter. Septal spines conical, short, sharp; occurring irregularly (Fig. 4E) but in some parts of coralla, they may be abundant.

Discussion. The above-described new species resembles the most *S. tschichatschewi* Peetz, from which it differs by much thinner walls (0.20–0.25 mm in axial, and up to 0.5 mm in peripheral zone of the *S. tschichatschewi*) and smaller connecting pores (0.2–0.3 mm in *S. tschichatschewi*). It is also similar to *S. tenuimuralis* Mironova, from which it differs by thicker walls in the peripheral part of the corallum (0.15 mm in *S. tenuimuralis*), strongly bent walls in the new species (flat in *S. tenuimuralis*) and finally different (horizontal in *S. tenuimuralis*) and less abundant tabulae. The wall thickness is considerably larger also in numerous other species (*S. khalfini* Dubatolov: 0.25–0.40 mm; *S. minuscula* Tchudinova:

0.20–0.40 mm; *S. kimi* Mironova: 0.10–0.25 mm). From *S. cerebra* Mironova, the new species differs by much smaller connecting pores (0.25–0.30 mm in *S. cerebra*), and from *S. illustra* Dubatolov by much smaller diameters of corallites (2.50–3.50 mm in *S. illustra*). *S.? enigmatica* Nowiński is a species with nearly all biometric parameters much smaller than the new species; moreover, by the arrangement of corallites, it resembles strongly members of the genus *Gracilopora*. Data concerning discussed species follow Dubatolov (1959, 1963), Tchudinova (1959, 1964), Mironova (1974) and Nowiński (1992).

Nowiński (1992) mentioned and illustrated *Hillaepora circulipora* Kayser from Laskowa Góra, Marzysz and Sowie Górki. I was not able to trace the material of *H. circulipora*; however, the sections from Marzysz were not signed (on the thin sections of A. Nowiński, there is usually a name of an organism and locality; here, the name was lacking). Most probably the new species of *Striatopora* is conspecific with Nowiński's *H. circulipora* (compare with Nowiński 1992, Fig. 4C–F). The status of the Kayser's species is unclear, as the original material is missing. Tourneur (1985, text-figs 181–182) however illustrated the topotypes, as well as material of *H. circulipora* from Belgium (Tourneur 1985, text-figs 180, 183–187). The material described above surely does not belong to the genus *Hillaepora*, neither does the material illustrated by Nowiński.

Occurrence. Kostomłoty Transitional Zone (Early–Middle Givetian), Central Kielce Subregion (Late Givetian).

Striatopora aff. peetzi Dubatolov, 1956
Figure 5C–E

Material. Kowala Railroad Cut; 19 fragments of coralla, 10 thin sections (ZPAL T.25: KOW R.006.1, KOW R.024.1–3, KOW R 102.1–2, KOW R.103.1–2, KOW R.105.1–2).

Description. Coralla branching, small, attain 10 mm in diameter, round or slightly oval in cross section. Calyces subrectangular, subpentagonal, with sharp edges, deep, arranged somewhat irregularly (Fig. 5C). They have lower wall bent, forming a lip. Corallites long, irregularly polygonal in the branch centre, with walls gradually bent in peripheral part of corallum. Corallites reach the corallum surface at acute angle close to 45 degrees. Corallite lumina polygonal in the centre of branch, becoming subtriangular or irregular in the peripheral part, often elongated. Diameters

of corallites variable, in the central parts of branch 0.54–0.70 × 0.80–0.90 mm in size, rarely to 1.30 mm of the largest diameter. Central and peripheral zones cannot be distinguished. Walls even, with moderate thickening towards the corallum surface. Their thickness in central part of the corallum varies from 0.06 to 0.26 mm, most often 0.16 to 0.20 mm. Median line thin, dark, continuous, well visible only in the axial part of branch. Tabulae rare, flat, concave, convex, twisted or strongly inclined, even forming dissepimenta, thin, spaced irregularly (spacing unmeasurable). Connecting pores very rare, 0.16–0.30 mm in diameter. Septal spines not observed.

Discussion. The specimens described here show strong affinity to *Striatopora peetzi* (specimen 81/134, the holotype, and paratypes, coll. V. Dubatolov, St. Petersburg State University) from which they differ by scarce connecting pores and their larger diameters, although the reliable measurements were taken only on two pores. Also branch diameter in the above-described species is somewhat larger; this may be, however, controlled environmentally.

Occurrence. Southern Kielce Subregion (Early Frasnian). *S. peetzi* was found in Russia (Kuznetsk Bassin and Peri-Salair) and the United States (Maryland); Early Devonian.

Striatopora? enigmatica Nowiński, 1992

*v. 1992 *Striatopora enigmatica* sp. n. Nowiński, p. 202, fig. 4A–B.
 v. 2003 *Striatopora enigmatica* Nowiński; Nowiński p. 133, pl. 79, figs 2–5.

Type material. The holotype is missing; paratypes are investigated here.

Material. Kowala (set A): Two poorly preserved specimens, four thin sections (ZPAL T.25: K032–033, K036–037).

Discussion. Nowiński (1992) tentatively assigned the species to *Striatopora*, but emphasized its unclear systematic position. Undoubtedly, several characters (such as strong distal thickening of the wall, well-marked axial zone) make it similar to certain species of *Gracilopora*, but until the status of the latter genus will not be resolved, the species is temporarily assigned to the genus *Striatopora*. This species has been illustrated by Nowiński (1992). I could not trace the type specimen; therefore, a neotype should

---

FIG. 4. A–E, *Striatopora sciuricauda* sp. nov., holotype. Laskowa Góra Quarry, Holy Cross Mountains, Poland; set A$_2$ Laskowa Beds, Middle Givetian. Specimen ZPAL T.25 LSG 001.1–4. A–B, transverse section. C, longitudinal section. D, tangential section. E, detail of Figure 3 (upper left part) showing strong, conical spines in one corallite. F, *Striatopora sciuricauda* sp. nov. Laskowa Góra Quarry, Holy Cross Mountains, Poland; set A (?), Laskowa Beds, Middle Givetian. External view of branching corallum. Specimen ZPAL T.25 LO.01. All scale bars represent 1 mm.

probably be established. The lectotype was not established here, as the material (Nowiński's paratypes) at author's disposal was of very poor quality.

Occurrence. Southern Kielce Subregion (Early Frasnian).

Striatopora? aff. *tenuis* Lecompte, 1939
Figure 5F–G

> ? 1952 *Striatopora* aff. *tenuis* Lecompte; Sokolov, p. 66, pl. 14, fig. 1.
> v 1992 *Striatopora tenuis* Lecompte; Nowiński, p. 190, fig. 3A–B.
> vp. 1992 *Striatopora* aff. *tenuis* Lecompte; Nowiński, p. 190, fig. 3C–D.
> v. 2003 *Striatopora tenuis* Lecompte; Nowiński, p. 134, pl. 91, fig. 1.

Material. Kowala (sets A–C): 22 fragments of coralla, four thin sections (ZPAL T.25 KOW 362–365); Laskowa Góra Quarry (set A): One corallum, two thin sections (ZPAL T.25 L008–009).

Description. Coralla branching. Branches slender, 3.5–7.5 mm in diameter. Calyces polygonal, very deep, up to 1.3 mm in diameter, forming lip. Corallites polygonal, inclined to the corallum surface at acute angle, usually between 45 and 60 degrees. Corallite dimensions very variable; in the axial zone, the corallite diameters vary between 0.34 and 0–58 mm. Corallite lumina usually round, sometimes feebly polygonal. Axial zone occupies about half of the branch diameter. Walls variable in thickness, 0.06–0.14 mm, even, thickened in corners, with faint distal thickening. Tabulae distributed very irregularly, thin, inclined, folded, sometimes even adhering to the wall, often absent on long sections of corallites. Connecting pores very rare, small, 0.10 mm in diameter. Septal apparatus not observed.

Discussion. The material described here shows remarkable similarities with *S. tenuis* described by Lecompte (1939) with two exceptions. In the type material, septal spines are present in the distal parts of corallites. Advanced weathering of investigated material may cause their absence in the material from Kowala. The second difference is the single wall thickness in the axial part of the branch (0.14–0.22 mm on the lectotype thin section MRHN a723, No. 444 from Couvin 8, Ardennes, Belgium; Lecompte, 1939, pl. 16, fig. 14). The attribution of Lecompte's material to the genus *Striatopora* is somewhat uncertain; the general shape of the corallites, that is, thick walls and rounded lumina, makes it similar to small species of *Gracilopora*. These forms are indeed very close, and the study on type population of *Striatopora tenuis* should be undertaken in order to recognize its affinities. The author's study on the type specimen could not resolve the problem because of too large thickness of Lecompte's thin section and faint obliquity of cutting. The material described above, as well as *S. tenuis* from the Ardennes, is similar to *S.? enigmatica* in the arrangement and shape of corallites. However, the diameters of the corallites of *S.?* aff. *enigmatica* are larger.

Occurrence. Kostomłoty Transitional Zone (Early Givetian), Southern Kielce Subregion (Early Frasnian), western slopes of Polar Urals (Givetian). *S. tenuis* has originally been described from the Eifelian of Belgium (Ardennes).

Genus THAMNOPORA Steininger, 1831

Type species. *Favosites cervicornis* (de Blainville, 1830) from the Middle Devonian(?) of Bensberg, Germany.

Diagnosis. Coralla branching, branches cylindrical, composed of corallites diverging from the centre of branch axis. Calyces opening to the branch surface at right or nearly right angle. Walls show angular thickening, so the corallite lumen is round or nearly round in cross section. Septal apparatus absent. Connecting pores numerous. Tabulae complete and flat. Based on the diagnosis of Tourneur 1986.

Discussion. Genus *Thamnopora* is one of the most common Devonian tabulate corals. Up to now, more than 100 species have been described from all over the world. One of the classical areas, providing numerous cosmopolitan species of *Thamnopora*, is the Ardennes region (Hubert *et al.* 2007). Lecompte (1939) described these thamnoporas in his seminal monograph, and Tourneur (1985) subsequently revised the material. The latter author redescribed most of the type specimens of Lecompte (1939) with additional material of his own. Tourneur (1985) stated that

1. Two kinds of microstructure occur in the genus *Thamnopora*: in one group of species the median line is composed of lamellae (*T. cervicornis*), while the

---

FIG. 5. A–B, *Striatopora sciuricauda* sp. nov. Marzysz, Holy Cross Mountains, Poland; Late Givetian. Specimen ZPAL T.25 M017–021. A, longitudinal section. B, transverse section. C, *Striatopora* aff. *peetzi* Dubatolov, 1956. Kowala Railroad Section, Holy Cross Mountains, Poland; rubble of set A, Early Frasnian. External view of branching corallum. Specimen ZPAL T.25 KOW RO.02. D–E, *Striatopora* aff. *peetzi* Dubatolov, 1956. Kowala Railroad Section, Holy Cross Mountains, Poland; rubble of set A, Early Frasnian. Specimen ZPAL T.25 KOW R.102.1–2. D, transverse section. E, longitudinal section. F–G, *Striatopora*? aff. *tenuis* Lecompte, 1939. Kowala Railroad Section, Holy Cross Mountains, Poland; rubble of set A, Early Frasnian. Specimen ZPAL T.25 K362–363. F, transverse section. G, longitudinal section. Scale bars represent 1 mm.

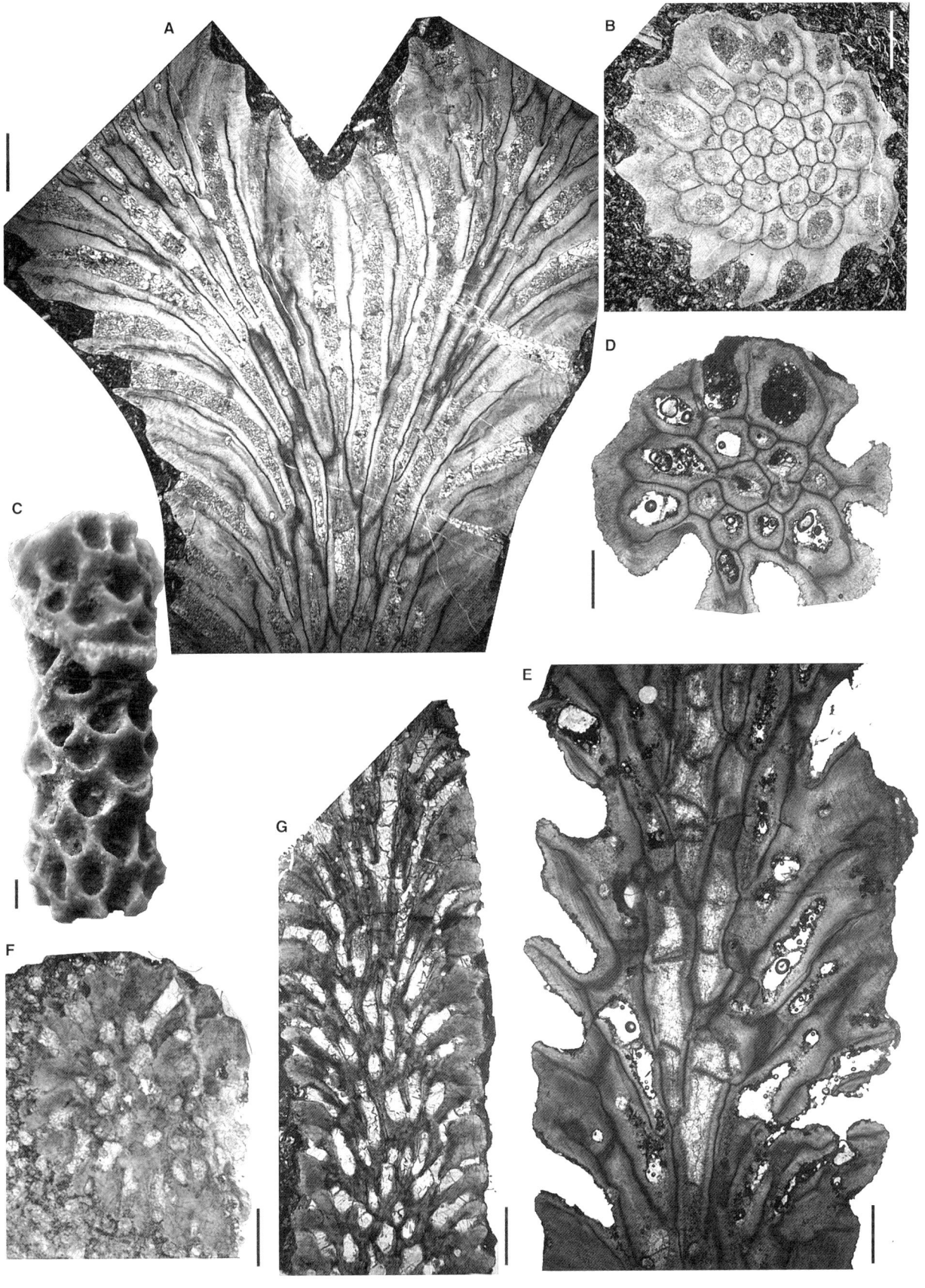

other group is characterized by granular microstructure of the median line (*T. tumefacta*); this observation led him to establish 1 year later a new genus, *Lecomptopora* Tourneur, with *T. tumefacta* as a type species (Tourneur 1986). In the present author's opinion, the difference in microstructure in only one skeletal element is insufficient for taxonomical purposes, as it may be diagenetic (see Oekentorp 2001 and discussion on microstructures above). Therefore, *Lecomptopora* is considered here as junior subjective synonym of *Thamnopora*.

2. The presence of the septal apparatus in the *Thamnopora* is problematic, as it is absent in the type species. Several species habitually assigned to *Thamnopora* show septal spines or squamulae (e.g. *T. proba*, *T. patula*, *T. auberti*). The presence of squamulae is often regarded as valid taxonomical feature in various tabulates (*Squameofavosites*, *Squameoalveolites*, *Caliapora*, see Tourneur 1985, p. 50; Iven *et al.* 1997). In a subsequent paper and in his revised diagnosis of *Thamnopora*, Tourneur (1986) stated that the septal apparatus is absent. Occurrence of septal apparatus can be controlled environmentally (Tourneur 1985, p. 50); the problem of generic attribution of species with septal spines has not been resolved. In contrast, the observations presented below on alveolitids permit the conclusion that development of a septal apparatus in this group (alveolitids) was controlled genetically, but that thamnoporids may have had different physiology.

The diagnoses of *Thamnopora* and *Lecomptopora*, given by Tourneur (1986, pp. 1255–1256), show that representatives of both genera have no septal spines. Thus, the systematic position of 'spiny' species remains open, and therefore, it may be therefore necessary to establish a separate genus. A new genus is not formally established here, as specimens being at the author's disposal do not show well-developed septal spines; it is necessary to investigate *Thamnopora* species with spines from Russia (that have much better developed septal apparatus than the Polish specimens).

May (1997, 1998) has published the most complete review of all known species of the genus. He found that some characters (e.g. diameter of the corallites in the axial part of branch vs. diameter in the peripheral part; branch diameter vs. pore diameter; corallite diameter in axial zone vs. peripheral diameter) are positively correlated, while others are not (e.g. branch diameter vs. pore spacing). The same author (May 1997, 1998) recognized *Gracilopora* Tchudinova as junior subjective synonym of *Thamnopora* on the basis of biometrical parameters. Morphometrical data are useful in taxonomy on specific and subspecific level; the generic assignment requires qualitative differences to exist (Imbrie 1956, p. 219).

On the basis of microstructural features, Plusquellec *et al.* (2007) excluded the genus *Thamnopora* from the family Pachyporidae; however, this viewpoint is not followed here owing to controversies in the application of microstructures in taxonomy.

Thamnopora cervicornis (de Blainville, 1830)
Figure 6E–J

*p	1830	*Alveolites cervicornis* de Blainville, p. 369.
non	1879	*Pachypora cervicornis* (de Blainville); Nicholson, p. 82, pl. 4, fig. 3.
?	1887	*Favosites (Pachypora) cervicornis* Blainville; Tschernychev, p. 121, pl. 4, fig. 23.
	1937	*Pachypora cervicornis* (Blainville); Tchernychev, p. 24, pl. 4, figs 2–3.
?	1952	*Thamnopora cervicornis* (de Blainville); Sokolov, p. 57, pl. 12, fig. 1–2, pl. 13, fig. 6.
?	1955	*Thamnopora cervicornis* (Blainville); Kraevskaya, p. 199, pl. 29, fig. 2.
?	1958	*Thamnopora boloniensis* (Gosselet); Stasińska, p. 198, pl. 10, fig. 1.
?	1959	*Thamnopora cervicornis* (de Blainville); Dubatolov, p. 101, pl. 32, fig. 1–5, pl. 33, fig. 1.
non	1959	*Thamnopora cervicornis* (de Blv.) var. *obtusispinosa* n. var.; Dubatolov, p. 102, pl. 33, fig. 2.
	1964	*Thamnopora cervicornis* (Blainville); Tchudinova, p. 43, pl. 19, figs 3–4, pl. 20, figs 1–2.
?	1972	*Thamnopora cervicornis* (Blainville); Yanet, p. 62, text-fig. 7, pl. 19, fig. 4.
	1985	*Thamnopora cervicornis* (de Blainville); Tourneur, p. 57, text-fig. 13–25, pl. 1–4. (*cum syn.*).
	1985	*Thamnopora cervicornis* (de Blainville); Birenheide, p. 70, text-fig. 3b (?), 17, pl. 18, fig. 1.
	1993*b*	*Thamnopora cervicornis* (de Blainville); May, p. 82, pl. 2, fig. 5–7.
vp.	2003	*Thamnopora cervicornis* (de Blainville); Nowiński, p. 135, pl. 73, fig. 1–3.
	2005	*Thamnopora cervicornis* (Blainville); Stadelmaier, Nose, May, Salerno, Schröder and Leinfelder, p. 14, pl. 5, fig. 1–7.

Type material. Specimen 259a (holotype), Geologisch-Paläontologisches Institut, Bonn, Germany, coll. A. Goldfuss.

Material. Kowala Railroad cut: 14 coralla 33 thin sections (ZPAL T.25: KoB3a, b; KoC4a, b; K230–231, K034–035, K001–004, K214–215, KOW 4.001.1, KOW R003–004, R018.1–3, R023.1–3, R043.1–4, R047.1–3, R048.1–3) from the rubble below the biostromal/biohermal complex; Sitkówka-Kostrzewa: four thin sections made probably of four coralla (ZPAL T.25 STA 97, 133, 134, 135).

Description. Coralla branching, round or oval in cross section, with branch diameter 8–15 mm. Axial zone occupies 1/2–2/3 of the branch diameter. Calyces polygonal, very deep, with rounded

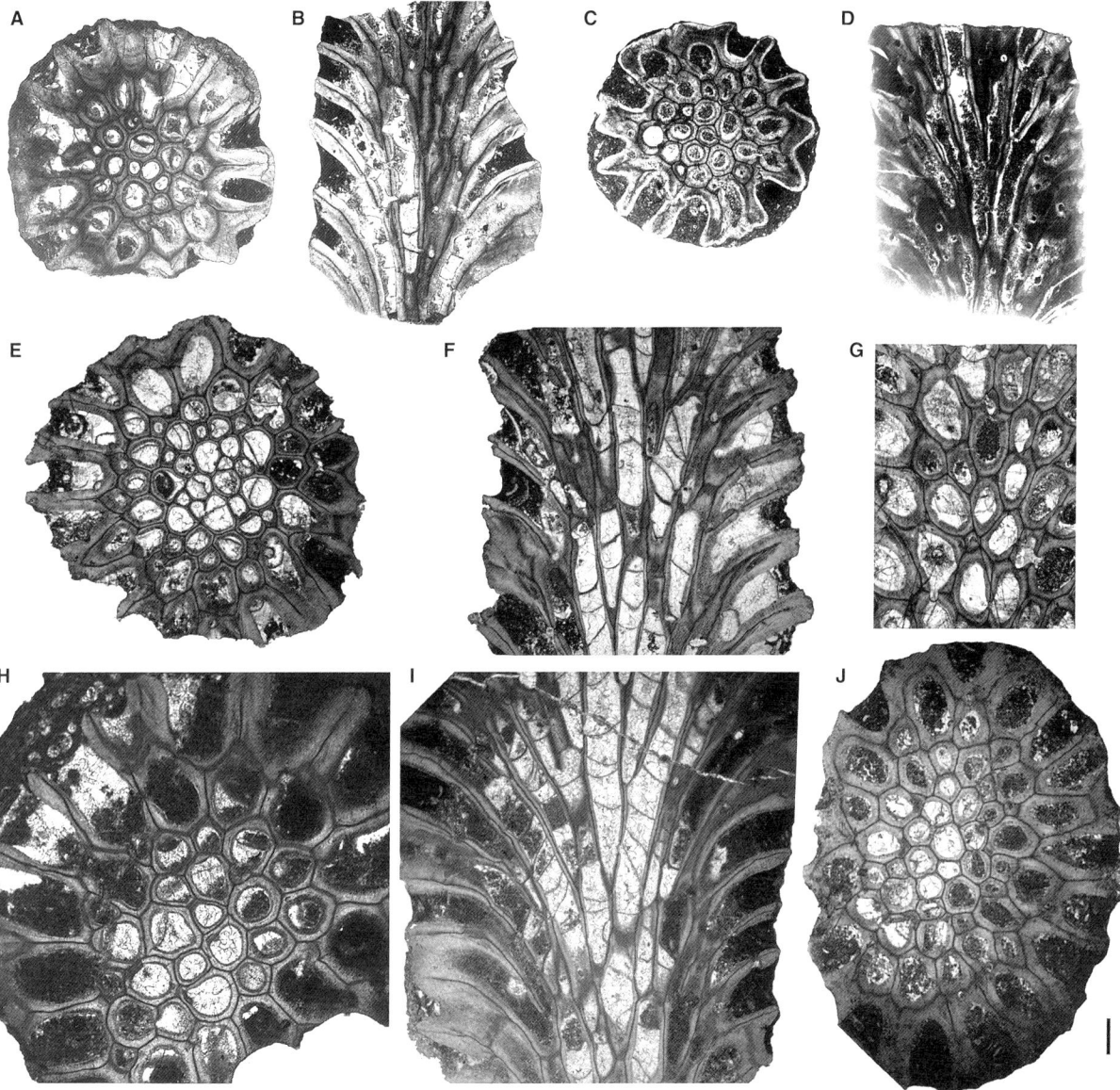

FIG. 6. A–D, *Thamnopora micropora* Lecompte, 1939. Kowala Railroad Section, Holy Cross Mountains, Poland; set A, Early Frasnian. Specimen ZPAL T.25 K281–282. A, transverse section. B, longitudinal section. Specimen ZPAL T.25 K242–243. C, transverse section. D, longitudinal section. E–J, *Thamnopora cervicornis* (de Blainville, 1830). Kowala Railroad Section, Holy Cross Mountains, Poland; rubble of sets A–C, Early Frasnian. Specimen ZPAL T.25 KOW R.047. E, transverse section. F, longitudinal section. G, tangential section. Specimen ZPAL T.25 KOW R.003-004. H, transverse section. I, longitudinal section. Specimen ZPAL T.25 KOW R.004. J, transverse section. Scale bars represent 1 mm.

edges, rather uniform in size. Axial zone markedly larger than peripheral, it occupies up to three-fourth of the cross section. Corallites polygonal, most often six-sided, up to eight-sided, long, reaching the surface of corallum at acute angle close to 90 degrees. In cross section, they are most often isometric, sometimes feebly elongated. Corallite lumina polygonal with rounded corners. Corallite diameter in axial zone varies from 0.74–1.30 × 0.80–1.80 mm, and in peripheral zone attains 2.30 mm. Walls thin, uneven, thickening in corners and towards the peripheral zone of corallum, 0.06–0.20 mm in thickness, most often 0.06–0.14 mm; in the peripheral zone, the wall thick-

ness may reach 0.30 mm. Stereoplasma seems to be composed of two layers, the inner one of radially arranged elements and the outer one probably concentrically arranged. In the axial part, the ratio between two layers is about 1:1, and in the peripheral part, the thickness of the outer layer strongly (relatively) decreases. The intermediate zone is somewhat blurred. In peripheral parts of corallum on longitudinal sections, clinogonal arrangement of crystals is visible. Median line thin, even, very well visible. Tabulae flat, inclined, sometimes strongly inclined, even infundibuliform, occasionally with faint foldings and irregularities. In distal parts of corallites, they may be slightly convex.

They are somewhat more abundant in the axial zone, spaced irregularly, 0.26–5.00 mm, most often 0.46–0.60 mm. They may be absent on long sections of corallites. Mural pores numerous, round, sometimes closed by poral plate, 0.20–0.30 mm, most often 0.20–0.26 mm in diameter. Septal apparatus absent.

Remarks. The specimens described here appear to be very similar to the lectotype described by Tourneur (1985, p. 60), except for the largest diameters of the corallites, which in the type material reach 1.60 mm. It must be stated that the value 2.30 mm is an extreme value obtained from a tangential section that may not be precisely parallel to the corallum surface. Also, the tabulae spacing in the material from the Holy Cross Mountains is somewhat larger. All other parameters (wall thickness, pore diameters, size of the axial zone) are similar to both the German and Belgian material.

Occurrence. Southern Kielce Subregion, Ardennes (Frasnian), Eifel Mts., Western slopes of Ural Mts. and Siberia (Givetian), Mongolia (Middle Devonian), Northern Spain (Upper Emsian – Lower Eifelian), probably also Kuznetsk Basin (Givetian).

Thamnopora micropora Lecompte, 1939
Figure 6A–D

    * v 1939   *Thamnopora micropora* nov. sp. *Lecompte*, p. 118, pl. 16, fig. 21
        1954   *Thamnopora micropora* Lecompte; Fontaine, p. 55, pl. 6, fig. 9–10.
        1987   *Thamnopora micropora* Lecompte; Boulvain, Coen-Aubert and Tourneur, p. 227, fig. 10.
  v p 2003   *Thamnopora micropora* Lecompte; Nowiński, p. 137, pl. 74, fig. 4–5, pl. 75, fig. 1.

Type material. Specimen MRHN 1558 (lectotype), Bruxelles, Belgium.

Material. Laskowa (set ?A): three coralla, six thin sections (ZPAL T.25: L043–044, L053–054, L059–060); Kowala Railroad Section (set C): five coralla, 11 thin sections (ZPAL T.25: K029–031, K242–243, K281–282, K343a, b), rubble: one corallum, two thin sections (ZPAL T.25 KQ013–014).

Description. Coralla branching, branches round in cross section, 7–9 mm in diameter. Calyces deep, with strongly rounded edges. Axial zone occupies slightly less than half of the branch diameter. Corallites polygonal, up to 9-sided, 0.62–1.1 × 0.74–1.40 mm in diameter. Corallite lumina strongly rounded, oval, round, subpolygonal. Walls even, 0.08–0.16 mm in axial zone, reaching some 0.30 mm near the corallum surface. Stereoplasma seems to be composed of two layers, the inner one of radially arranged elements and the outer one of probably concentrically arranged. In the axial part, the ratio between two layers is variable, often one of the zones is lacking. The median line is well visible on cross sections; on longitudinal sections, it often becomes blurry. Tabulae very variable in thickness, rare, spaced every 0.60–0.90 mm. They are flat, inclined, folded, rarely collapsed. Connecting pores frequent, round, often closed by poral plate, rather uniform in diameter (0.10 mm), spaced 0.60–0.90 mm. Septal apparatus absent.

Discussion. The material described above is very similar to the type specimen of Lecompte (MRHN 1558, thin section a498 from Senzeille 7153, Ardennes, Belgium; Frasnian). In the type material, the tabulae are somewhat more frequent and the two layers of stereoplasma are not as well visible as in the Holy Cross Mountains material; moreover, the axial zone is feebly larger than in the Polish material. It also needs to be noted that in the type material, the corallite walls are very variable in thickness; this feature is not observed in the Polish material. The type material requires certainly a revision, as it was not covered by Tourneur's (1985) monograph. The determination of the material from Laskowa is somewhat doubtful, owing to different shape and abundance of tabulae (flat and very frequent).

Occurrence. Northern Kielce Subregion (Early–Middle Givetian), Southern Kielce Subregion (Early Frasnian), West Germany (Frasnian), Laos (Frasnian?).

Thamnopora ex gr. *boloniensis* (Gosselet, 1877)
Figure 7A–I

        1840   *Alveolites cervicornis* Blainville; Michelin, p. 187, pl. 48, fig. 2, pl. 49, fig. 3.
  * ? 1877   *Favosites boloniensis*; Gosselet, p. 271
        1880   *Favosites boloniensis*; Gosselet, p. 53, (?) pl. 4, fig. 22.
   v. 1939   *Thamnopora boloniensis* (Gosselet); Lecompte, p. 122, pl. 17, fig. 1.
   p 1958   *Thamnopora boloniensis* (Gosselet); Stasińska, p. 198, pl. 9, fig. 1–4, pl. 10, fig. 2–4, pl. 11, fig. 1–2. non pl. 10, fig. 1.

FIG. 7. A–I, *Thamnopora* ex gr. *boloniensis* (Gosselet, 1877). Kowala Railroad Section, Holy Cross Mountains, Poland; rubble of sets A–C, Early Frasnian. Specimen ZPAL T.25 KOW R.016. A, transverse section. B, longitudinal section. C, tangential section. Specimen ZPAL T.25 KOW R.013. D, transverse section. E, longitudinal section. F, tangential section. Specimen ZPAL T.25 KOW R.046. G, transverse section. H, longitudinal section. I, tangential section. J–K, *Thamnopora* cf. *irregularis* Lecompte, 1939. Jaźwica Quarry, Holy Cross Mountains, Poland; set H, Early Frasnian. Specimen ZPAL T.25 J016–017. J, transverse section. K, longitudinal section. L–M, *Thamnopora* sp. A. Laskowa Góra Quarry, Holy Cross Mountains, Poland; Laskowa Beds, Givetian. Specimen ZPAL T.25 L057–058. L, transverse section. M, longitudinal section. Scale bars represent 1 mm.

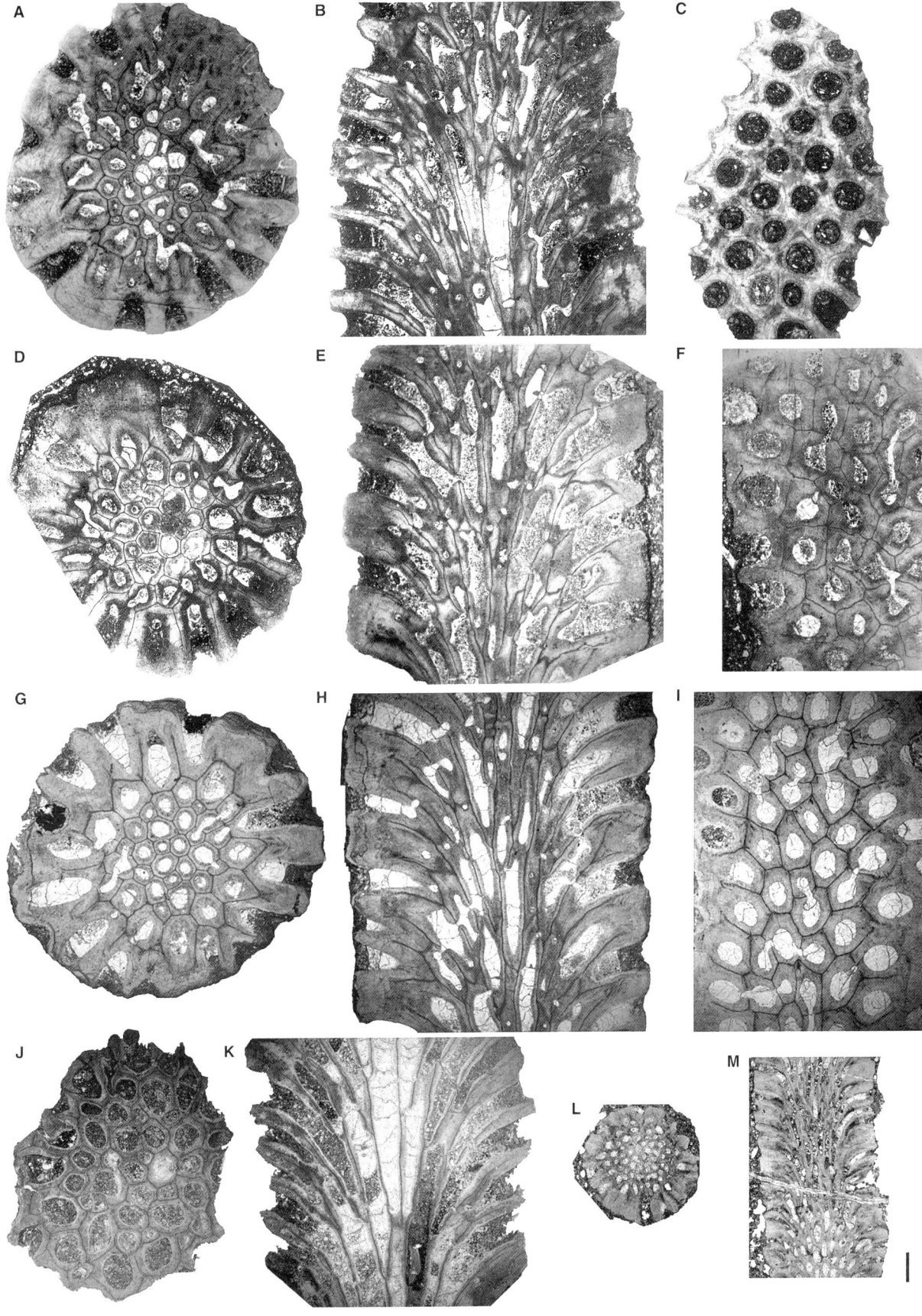

? 1959   *Thamnopora boloniensis* (Gosselet); Dubatolov, p. 111, pl. 29, fig. 2.

1976   *Thamnopora boloniensis* (Gosselet); Nowiński, p. 53, pl. 5, fig. 3–8.

? 1984a   *Thamnopora boloniensis boloniensis* (Gosselet); Hladil, p. 36, pl. 2, fig. 6.

? 1984a   *Thamnopora boloniensis langi* subsp. n. Hladil, p. 37, pl. II, fig. 4–5.

1985   *Thamnopora boloniensis* (Gosselet); Tourneur, p. 262, text-fig. 144–150 (*cum syn.*).

1988   *Thamnopora boloniensis* (Gosselet); Tong-Dzuy, Nguyen and Kromykh, p. 81, pl. 36, fig. 1.

1995   *Thamnopora boloniensis* (Gosselet); May, p. 484, pl. 2, fig. 9–11.

v. 2003   *Thamnopora boloniensis* (Gosselet); Nowiński, p. 135, pl. 70, fig. 1–3, pl. 72, fig. 3–4.

Material. Bilcza (set B): Two coralla, six thin sections (ZPAL T.25 BI001–006); Czarnów (set A): Four thin sections made of three coralla (ZPAL T.25 C009, C012–013, C028–029); Jaźwica (set H): two thin sections made of one corallum (ZPAL T.25 J018–019); Kowala Railroad Section and Quarry (sets A, C): About 90 fragments of coralla (biohermal complex and rubble), 162 thin sections (ZPAL T.25: KOW 4.001.1–2, KOW 9.001.1–5, KOW R.002, KOW R.005.1–2, KOW R.014.1–3, KOW R.017.1–2, KOW R.019.1–3, KOW R.020.1–3, KOW R.021.1–2, KOW R.022.1–4, KOW R.025.1–3, KOW R.026.1–3, KOW R.038.1–3, KOW R.039.1–3, KOW R.041.1–3, KOW R.042.1–3, KOW R.045.1–3, KOW R.046.1–3, KOW R.049.1–3, KOW R.050.1–3, KOW R.051.1–3; ZPAL T.25 K021–022, K038, K046–063, K068–075, K083–085, K091–093, K097, K099–105, K107–110, K116–121, K123–128, K131–133, K149–152, K162–165, K167–171, K174, K203–208, K216, K224a–b, K235–236, K239–241, K248–251, K272–274, K430–432), and 16 thin sections made of unknown number of coralla (ZPAL T.25: STA 74–86, STA 125, STA 127, STA 132); Laskowa Góra Quarry: two thin sections made of one corallum (ZPAL T.25 L063–064); Sitkówka–Kostrzewa: one specimen (several branches), 10 thin sections (ZPAL T.25 SIT 13.001.1–10); Sowie Górki (sets B, C): two coralla, six thin sections (ZPAL T.25 SG061–066).

Description. Coralla branching with branch diameter 8–12 mm. Calyces polygonal, deep, with rounded edges, rather uniform in size. Axial zone markedly larger than peripheral. Corallites polygonal, most often six-sided, up to seven-sided, long, reaching the surface of corallum at acute angle close to 90 degrees. In cross section, they are isometric or elongated. Corallite lumina round or strongly oval. Corallite diameter in axial zone varies from $0.80–1.60 \times 0.82–1.66$ mm and in peripheral zone attains 2.00 mm. Walls thick, strongly thickening in corners and towards the peripheral zone of corallum, 0.08–0.28 mm in thickness, most often 0.10–0.22 mm. Stereoplasma seems to be composed of two layers, the inner one of radially arranged elements, the outer one of probably concentrically arranged. In the axial part, the ratio between two layers is about 1:1, and in the peripheral part, the thickness of the inner layer strongly decreases. The intermediate zone is somewhat blurry. In peripheral parts of corallum on longitudinal sections, clinogo-

nal arrangement of crystals is visible. Median line thin, even, very well visible. Tabulae flat, inclined, sometimes strongly inclined, occasionally with faint foldings and irregularities, slightly more abundant in the axial zone, spaced irregularly, 0.38–2.00 mm. Mural pores numerous, round, sometimes closed by poral plate, 0.14–0.30 mm in diameter, spaced 0.86–2.00 mm. They are smaller in axial zone, becoming feebly larger towards periphery of the branch. Septal apparatus generally absent, but in few specimens from Kowala small, spine-like structures are visible.

Discussion. The type specimen of the discussed species is this figured by Michelin (1840, pl. 48, fig. 2) and designated by Gosselet (1877). The type specimen is untraceable in the collections of the Muséum national d'Histoire naturelle in Paris, and for that reason, a neotype needs to be chosen (see Tourneur 1985, p. 263). Dubatolov (1959) and Hladil (1984a) have chosen lectotypes, but they are invalid, as neither they were chosen from the syntypes (art. 74.2 of the ICZN); nor they come from the type locality (original material of Gosselet comes from Boulonnais, France, while both lectotypes were chosen from the material from Ardennes, Belgium). For further comments concerning choice of lectotypes, see Remarks chapter in *Alveolites parvus* description.

Tourneur (1985, p. 255) in the description of *Th. polyforata* Schlottheim stated that this species is badly defined, mainly due to the large intraspecific variation of type population, and stated that both *Th. polyforata* and *Th. boloniensis* are closely related species. In the present author's opinion, they may even be conspecific. Their affinities can be recognized only after reinvestigation of the type population of *Th. boloniensis*. Consequently, as the species is not well defined, the specimens described here are assigned to the *Thamnopora boloniensis* species group.

The species seems to be very common and has been reported from all over the world, but until the neotype will not be established, it will be impossible to define precisely its characters and distribution; therefore, all determinations will require revision in future.

Occurrence. Kostomoty Transitional Zone (Givetian), Central and Southern Kielce Subregions (Givetian–Early Frasnian); Boulonnais (France, Givetian); Ardennes (Givetian–Frasnian), Eifel Mts. (Givetian), Viet Nam (Frasnian), Kuznetsk Basin (?; Frasnian); Moravia (?; Frasnian); Western Pomerania (?, Poland, Frasnian).

Thamnopora cf. *irregularis* Lecompte, 1939
Figure 7J–K

* v ? 1939   *Thamnopora irregularis* nov. sp. Lecompte, p. 113, pl. 15, fig. 6–9.

? 1985 *Thamnopora irregularis* Lecompte; Tourneur, p. 107, figs 45–50.
? 1993*b* *Thamnopora irregularis* Lecompte; May, p. 80, pl. 1, fig. 5–12.
v. 2003 *Thamnopora irregularis* Lecompte; Nowiński, p. 136, pl. 74, fig. 2–3.
? 2005 *Thamnopora irregularis* Lecompte; Stadelmaier, Nose, May, Salerno, Schröder and Leinfelder, p. 16, pl. 6, fig. 1–4.

Type material. Specimen MRHN 6151a/264 (lectotype), Bruxelles, Belgium.

Material. Kowala Railroad Section (set ?A): one corallum, three thin sections (ZPAL T.25 K258–260); Jaźwica (set H): one corallum, two thin sections (ZPAL T.25 J016–017).

Description. Coralla branching, branches round in cross section, 9–10 mm in diameter. Calyces deep, edges not preserved. Axial zone wide, occupies more than ¼ of the branch diameter. Corallites polygonal, up to 9-sided, measuring 0.84–1.34 × 1.16–1.86 in diameter. Corallite lumina feebly rounded, subpolygonal. Walls even, 0.08–0.20 mm in axial zone, with faint distal thickening. The median line is thin, but well visible. Tabulae are very thin, spaced every 0.34–0.80 mm. They are flat, inclined, folded, sometimes with umbilical pits. Connecting pores are frequent, rounded and often closed by poral plate, 0.20–0.30 mm in diameter, spacing not measurable. Septal apparatus absent.

Remarks. The material is very similar to the type material of Lecompte, except for the more uniform size of corallites, which may depend on budding frequency (Tourneur 1985, p. 111). Also, pore diameters in the Polish specimens are feebly larger.

Occurrence. Southern Kielce Subregion (Early Frasnian), probably Western Pomerania (Givetian), Ardennes (Givetian).

Thamnopora sp. A
Figure 7L–M

? 1952 *Gracilopora vermicularis* (McCoy); Sokolov, p. 71, pl. 15, fig. 1.
v. 2003 *Gracilopora vermicularis* (McCoy); Nowiński, p. 132, pl. 80, fig. 1.

Material. Jaźwica/Góra Łgawa: One corallum, two thin sections (ZPAL T.25 J059–060); Laskowa (set A or B): four coralla, seven thin sections (ZPAL T.25: L057–058, L061–062, L120–122); Posłowice: one corallum, one thin section (ZPAL T.25 P110–20).

Description. Coralla very small. Branches slender, usually 3–4 mm in diameter. Calyces deep, with rounded edges. The axial zone occupies about three-fourth of the branch diameter. Corallites long, prismatic, reaching the surface at acute, nearly right angle. Corallites in the axial zone polygonal with rounded corners, 0.26–0.40 mm in diameter in the axial zone; lumina subpolygonal in outline. Walls thin in axial zone, thickening strongly in the peripheral zone. Median line blurry, not well visible in the axial zone; it becomes invisible towards the periphery of the corallum. In the axial zone, single wall thickness varies between 0.04 and 0.06 mm. Tabulae flat or inclined, very rare. Connecting pores rare, round, about 0.10 mm in diameter. Septal spines not observed.

Discussion. The specimens analysed here show most of characters typical for *Thamnopora*, but Nowiński (1992, 2003) described and referred the material to *G. vermicularis* (McCoy, 1850). The latter species is badly defined; moreover, it was not illustrated in the original paper, and until now, it has not been revised. May (1997) recognized the genus *Gracilopora* as a junior subjective synonym of *Thamnopora*. Not regarding its validity, these specimens seem to be close to the genus *Thamnopora*.

Biometrical characters of the discussed species make it close to two thamnoporids: *Gracilopora acuta* Tchudinova, 1964 and *Thamnopora longdongshuiensis* Deng, 1979 (for data on biometry see May 1997). These two species share most of characters with *Thamnopora* sp., and they differ in having very weakly developed septal spines. Moreover, these species were described from the Eifelian of Asia; therefore, time and space gap does not allow to assign the described here specimens to either of the species.

Occurrence. Kostomłoty Transitional Zone (?Early–Middle Givetian), Central Kielce Subregion (Late Givetian), Southern Kielce Subregion (?Late Givetian–Frasnian).

Suborder ALVEOLITINA Sokolov, 1950
Family ALVEOLITIDAE Duncan, 1872

Subfamily ALVEOLITINAE Duncan, 1872

Discussion. Several genera belonging to this subfamily seem to be easily recognizable, having clear features, and others are vaguely distinct. Hladil (1981*a*, pp. 28–29) highlighted the need of revision of *Alveolites* and related genera, owing to their unclear definitions. The main features of closely genera belonging to the discussed subfamily are given below:

Alveolites Lamarck, 1801 has massive or tabular coralla, connecting pores in corners, usually flattened corallites. Septal apparatus may be developed to various degree (diagnosis by Hill 1981).

Alveolitella Sokolov, 1952 has small, branching coralla with axial zone clearly marked, while *Alveolites* has medium or large, massive or tabular ones (diagnosis by May

1993a); *Crassialveolites* Sokolov, 1955 has very thick coral-lite walls, with nearly round lumen (diagnosis by Sokolov 1962a).

Grandalveolites Mironova, 1970 is probably a junior synonym of *Alveolites*, as there is no clear feature allowing distinguishing these genera (diagnosis by Hill 1981).

Gokaselites Niko and Adachi, 1999 differs from *Crassi-alveolites* by encrusting corallum and connecting protu-berances (diagnosis by Niko and Adachi 1999).

Kitakamiia Sugiyama, 1940 has a chevron-shaped coral-lites with a large septal spine on basal lip (diagnosis by Hill 1981).

Planalveolites Lang and Smith, 1939 has very flat, thin coralla and thin-walled corallites (see Stumm 1967 for diagnosis; according to Lafuste (1984) and Plusquellec *et al.* (1997) this genus should be placed in Tabulata as *Incertae sedis*).

Scharkovaelites Mironova, 1974 has pores at faces of walls and in the corners. Such a feature in the present writer's opinion poses the representatives of this genus between Favositinae and Alveolitinae (diagnosis by Hill 1981).

Subalveolitella Sokolov, 1955 has sharp differentiation of corallites: thin-walled, polygonal in the axial zone of branch, compacted and thick-walled in the peripheral part (diagnosis by Sokolov 1962a, b).

Subalveolites Sokolov, 1955 has very thin walls and large visceral cavity in relation to the width of wall (diag-nosis by Sokolov 1962a).

The anatomy of alveolitids strongly depends on envi-ronmental conditions. An exemplifying study was pre-sented by Hladil (1989a).

Genus ALVEOLITES Lamarck, 1801

Type species. *Alveolites suborbicularis* from the Middle–Late Devonian of Bensberg (Germany).

Remarks. Specimens having branched coralla belong to the genus *Alveolitella* Sokolov. Therefore, information on branching coralla given by Hill (1981) should be removed from the diagnosis.

Alveolites suborbicularis Lamarck, 1801
Figure 8A–F

*	1801	*Alveolites suborbicularis* Lamarck, p. 186.
	1851	*Alveolites suborbicularis* Lamarck; Milne-Edwards and Haime, pp. 153, 255.
	1879	*Alveolites suborbicularis* Lamarck; Nicholson, p. 126, text-fig. 20, pl. 6, fig. 2.
?	1887	*Alveolites suborbicularis* Lamarck; Tscherny-chev, p. 122, pl. 4, fig. 24.
	1904	*Alveolites suborbicularis* Lamarck; Sobolev, p. 27, pl. 3, fig. 5.
non	1908	*Alveolites suborbicularis* Lamarck; Reed, p. 20, pl. 4, figs 3–4.
? v	1933	*Alveolites labechei* Milne-Edwards and Haime; Lecompte, p. 25, pl. 1, figs 3–4.
	1933	*Alveolites suborbicularis* Lamarck; Lecompte, p. 15, pl. 1, figs 1–2.
	1933	*Alveolites suborbicularis* Lamarck; Smith, p. 137, pl. 2, figs 2–3.
	1936	*Alveolites suborbicularis* Lamarck; Lecompte, p. 6, pl. 1, figs 1–2, pl. 2, figs 1–2.
p v	1939	*Alveolites suborbicularis* Lamarck; Lecompte, p. 19, pl. 1, figs 1–18.
p v	1939	*Alveolites suborbicularis* forma *gemmans* Leco-mpte, p. 22, pl. 1, figs 1–12.
? v	1939	*Alveolites suborbicularis* forma *subramosa* Lecompte, p. 23, pl. 1, figs 13–16.
non v	1939	*Alveolites suborbicularis* forma *contorta* Leco-mpte, p. 23, pl. 1, figs 17–18.
? v	1939	*Alveolites suborbicularis* var. *lamellosus* n. var. Lecompte, p. 24, pl. 2, fig. 3.
	1939	*Alveolites suborbicularis* Lamarck; Kelus, p. 46, fig. 39a, b.
	1952	*Alveolites suborbicularis* Lamarck; Sokolov, p. 78, pl. 17, figs 1–4.
?	1952	*Alveolites suborbicularis* Lamarck var. *minor* Frech; Sokolov, p. 80, pl. 18, figs 3–6.
non	1952	*A. suborbicularis* Lamarck var. *lamellosa* Leco-mpte; Sokolov, p. 81, pl. 18, figs 1–2, pl. 19, figs 1–2.
p v	1953	*Alveolites suborbicularis* Lamarck; Stasińska, p. 232, text-figs 12, 13, pl. 4, figs 1–3.
?	1954	*Alveolites suborbicularis* Lamarck; Fontaine, p. 26, pl. 1, fig. 3.
	1959	*Alveolites suborbicularis* Lamarck; Dubatolov p. 142; pl. 47, fig. 4A–G.

FIG. 8. A–C, *Alveolites suborbicularis* Lamarck, 1801. Jaźwica Quarry, Holy Cross Mountains, Poland; set K (?), Early Frasnian. Specimen ZPAL T.25 JAZ 005. A, transverse section. B, longitudinal section. Specimen ZPAL T.25 JAZ 009. C. longitudinal section. Arrow shows connecting pore closed by poral plate. D–F, *Alveolites suborbicularis* Lamarck, 1801. Kowala Railroad Section, Holy Cross Mountains, Poland; set A, Early Frasnian. Specimen ZPAL T.25 KOW 4.001. D, transverse section. E, longitudinal section. F, transverse section. G–L, *Alveolites compressus* (Milne-Edwards and Haime, 1853). Kowala Railroad Section, Holy Cross Mountains, Poland; set A, Early Frasnian. Specimen ZPAL T.25 KOW AX.001. G, transverse section. H, longitudinal section. I, transverse section. J, transverse section. Photographs I and J are from acetate peels and show the differences in double wall thickness. Peels are spaced 1.4 mm apart. Specimen ZPAL T.25 KOW 12.001. K, transverse section. L, longitudinal section. Scale bars represent 1 mm.

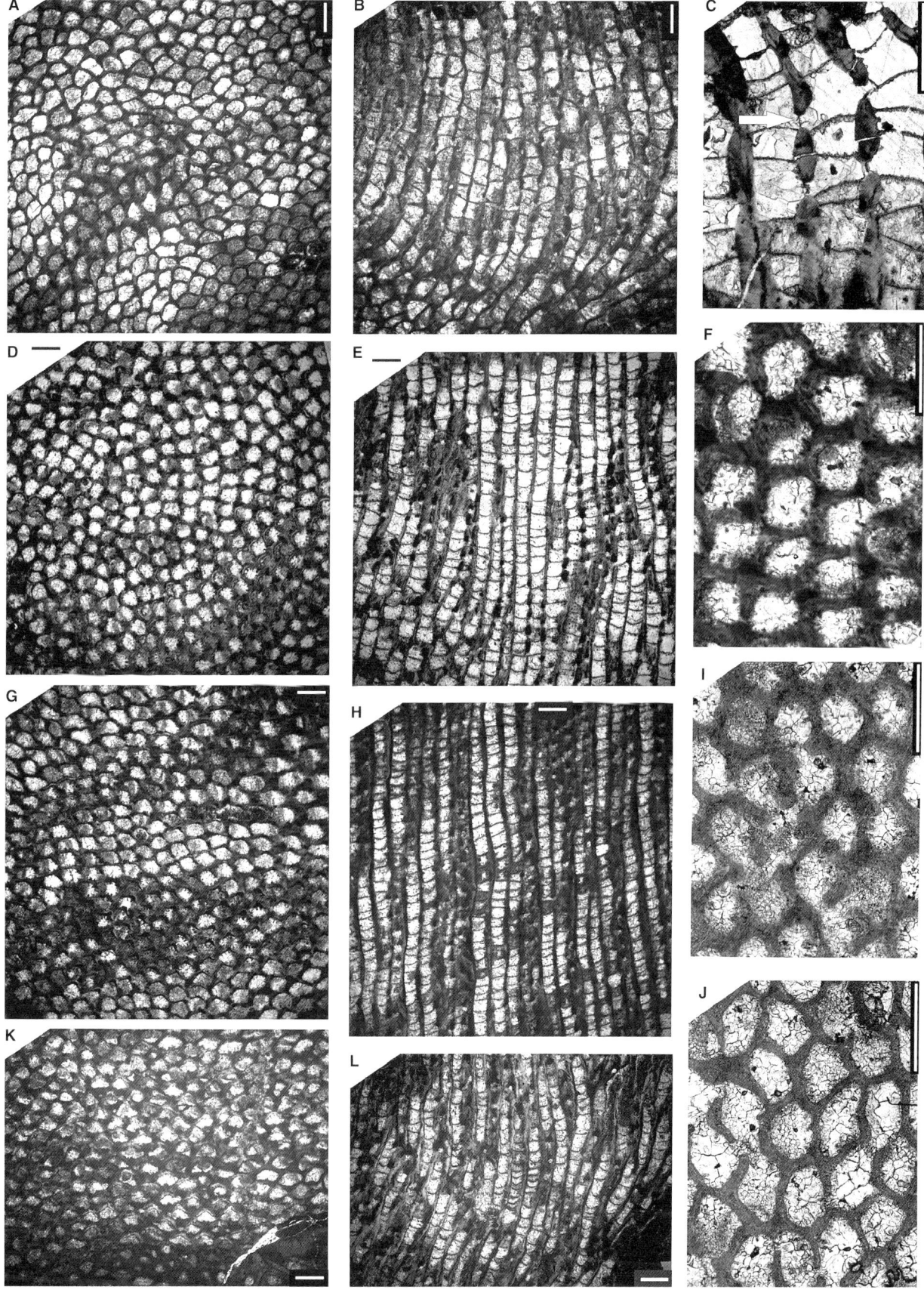

1976   *Alveolites suborbicularis* Lamarck; Nowiński, p. 59, pl. 9, figs 1–2.
1980   *Alveolites suborbicularis* Lamarck; Iven, p. 136, pl. 1, fig. 6–9; pl 2, figs 1–7.
1980   *Alveolites pseudorbicularis* nov. sp. Iven, p. 137, pl. 3, figs 1–3.
1985   *Alveolites suborbicularis* Lamarck; Birenheide, p. 79, pl. 21, fig. 1.
1993a   *Alveolites (Alveolites) suborbicularis* Lamarck; May, p. 145.
1996   *Alveolites suborbicularis* Lamarck; Brühl, p. 10, pl. 1, figs 1–2, pl. 2, fig. 3.
? 1998 *Alveolites suborbicularis* Lamarck; Birenheide, p. 182, pl. 2, fig. 3.
? 1999 *Alveolites suborbicularis* Lamarck; Brühl and Pohler, p. 279, pl. 4, figs 1–3.
v. 2003 *Alveolites suborbicularis* Lamarck; Nowiński, p. 144, pl. 87, fig. 3.

Type material. Specimen in A. Goldfuss collection (lectotype; see Goldfuss 1829, pl. 28, fig. 1a–b and Smith 1933, p. 138), Bonn University, Germany.

Material. Czarnów (set B): two coralla, five thin sections (ZPAL T.25: C006–008, C031–032); Góra Cmentarna: two coralla, four thin sections (ZPAL T.25: GC 001–003, GC 007–008); Jaźwica (sets I–K): 10 coralla, 25 thin sections (ZPAL T.25: J009–011, J021–023, J068–070; JAZ 005.1–2, JAZ 006.1–3, JAZ 007.1–3, JAZ 009.1–3, JAZ 023, JAZ 027.1–2, JAZ 029.1–2); Kadzielnia (set A): two thin sections (ZPAL T.25: STA KA 002–003); Kowala (sets A–C): 34 coralla, 97 thin sections (ZPAL T.25: KOW 4.001.1–5, KOW 8.001.1–2, KOW 8.002.1–3, KOW 9.002.1–3, KOW 9.A01.1–5, KOW 11.003.1–2. KOW R.A02.1–2, KOW R.041.1–3, KOW R.060, KOW R.061.1–2, K094–095, K134–148, K153–157, K209–216, K219–223, K225–226, K269, K275–276, K289–292, K301–311, K360–361); Laskowa (set A): one corallum, two thin sections (ZPAL T.25 L071–072); Miedzianka: one corallum, two thin sections (ZPAL T.25 MIE W.004.1–2); Posłowice: one corallum, three thin sections (ZPAL T.25 PW 050–052); Psie Górki: six thin sections (ZPAL T.25 STA PG 001–004); Wietrznia: two thin sections (ZPAL T.25 STA W 009–010).

Description. Coralla massive, nodular aberrant, medium and large, measuring up to 150 mm of the largest diameter. Calyces shallow with polygonal rounded outline and rounded edges. Corallites long, rounded-prismatic, slightly curved or feebly meandering. Cross section of corallites highly varies in shape: alveolitid, alveolitid-elongated, subrhomboidal, subrectangular, subtriangular, rarely polygonal and irregular. Lumina with rounded angles, variable in size (see table below). Walls uneven, with median line usually poorly preserved, visible as discontinuous dark line or crescentic shape on wider side of corallite. Connecting pores generally not frequent, but in some places within coralla, may be abundant. They are round, with variable spacing. Poral plates may be present; they occur accidentally throughout corallum. Tabulae numerous, complete, flat, slightly concave or convex, rarely folded, zones with numerous incomplete tabulae

may be present; they are distributed irregularly. Septal apparatus variously developed different places of corallum – from absent or nearly absent, through small button-like, blunt, rare spines, to (rarely) crown of short, sharp spines.

Measurements. Table 1.

Discussion. Alveolites suborbicularis, type species of the genus *Alveolites*, is one of the most commonly 'recognized' Devonian taxa. As it was already noticed by Stasińska (1953), *Alveolites suborbicularis* is one of the most common species in the Frasnian of the Holy Cross Mountains. This species is often misdetermined or treated as 'collective species', where all specimens, at least resembling it, are assigned. In fact, since the papers by Smith (1933) and Lecompte (1936, 1939), this species was not revised. Following the descriptions by Smith (1933) and Lecompte (1936), it can be stated that the described here material is very similar to the lectotype. The only difference may be the maximal lumen values, smaller in the analysed material, but not knowing how the measurements by Smith were taken, it cannot be stated whether this difference is significant.

The intracolonial variation of *A. suborbicularis* is similar to other species of the genus. In some specimens (e.g. ZPAL T.25 KOW 4.001), the variation of corallite lumina is very low, unlike in most of the representatives of the genus. The growth pattern of this species was described by Zapalski *et al.* (2012).

Occurrence. Kostomłoty Transitional Zone (Early Givetian), Kielce Region (Middle Givetian–?Late Frasnian), Łysogóry Region (Eifelian–Givetian), Boulonnais (France; Frasnian), Ardennes (Givetian–Frasnian), Eifel Mts. (Eifelian), Vestfalia (Frasnian), Main Devonian Field, Central Devonian Field (Frasnian), Kuzbas, Altai, Western Periuralia, Ural Mts. (Middle Devonian); Volhynia (Middle Devonian), Yunnan (Eifelian), Australia (late Eifelian – early Givetian).

Alveolites compressus Milne-Edwards and Haime, 1853
Figure 8F–L

* 1853 *Alveolites compressa* Milne-Edwards and Haime, p. 221, pl. 49, fig. 3.
v. 1933 *Alveolites compressa* Milne-Edwards and Haime; Lecompte, p. 27, pl. 1, figs 5–6.
v. 1992 *Alveolites compressus* Milne-Edwards and Haime; Nowiński, p. 192, text-fig. 5.
v. 2003 *Alveolites compressus* Milne-Edwards and Haime; Nowiński, p. 140, pl. 82, figs 1–2.
vp 2003 *Alveolites complanatus* Lecompte; Nowiński, p. 140, pl. 81, figs 2–3.

Type material. Unknown; see Remarks.

TABLE 1. Intracolonial variation in *Alveolites suborbicularis* Lamarck.

	Max LD	Min LD	DWT	PD	PS	TS
Specimen ZPAL T.25 KOW R.060, Kowala Quarry, rubble, Early Frasnian						
Mean	0.671	0.445	0.137	0.149	0.728	0.714
Standard deviation	0.1166	0.0749	0.0515	0.0188	0.1369	0.2318
N	51	51	51	33	15	38
Minimal value	0.50	0.30	0.06	0.12	0.44	0.22
Maximal value	0.96	0.74	0.32	0.20	1.04	1.20
V	0.174	0.168	0.376	0.126	0.188	0.325
Specimen ZPAL T.25 KOW 4.001, Kowala Railroad section, biohermal complex, Early Frasnian						
Mean	0.657	0.537	0.162	0.151	0.547	0.506
Standard deviation	0.0614	0.0653	0.0334	0.0196	0.1428	0.1342
N	35	35	35	35	35	60
Minimal value	0.50	0.42	0.10	0.10	0.36	0.24
Maximal value	0.80	0.70	0.22	0.20	1.00	0.76
V	0.093	0.122	0.206	0.129	0.261	0.265
Specimen ZPAL T.25 KOW 9.A01, Kowala Railroad section, set A, Early Frasnian						
Mean	0.620	0.420	0.167	0.18	–	0.553
Standard deviation	0.0795	0.0509	0.0304	0.0183	–	0.1902
N	30	30	30	13	7	30
Minimal value	0.48	0.32	0.12	0.16	0.46	0.14
Maximal value	0.8	0.52	0.22	0.20	0.88	1.00
V	0.128	0.121	0.182	0.101	–	0.344
Specimen ZPAL T.25 J009–011, Jaźwica, G. Łgawa, set C, Late Givetian						
Mean	0.663	0.466	0.165	0.154	–	0.432
Standard deviation	0.1151	0.0804	0.0454	0.0269	–	0.1556
N	51	51	52	52	6	52
Minimal value	0.5	0.34	0.06	0.10	0.30	0.18
Maximal value	0.92	0.68	0.26	0.22	0.58	0.94
V	0.174	0.173	0.276	0.174	–	0.360
Specimen ZPAL T.25 JAZ 006, from Jazwica, Frasnian						
Mean	0.598	0.458	0.134	–	–	0.587
Standard deviation	0.0679	0.0675	0.0357	–	–	0.0997
N	32	32	32	–	–	11
Minimal value	0.4	0.26	0.08	–	–	0.48
Maximal value	0.74	0.58	0.20	–	–	0.74
V	0.114	0.147	0.267	–	–	0.170
Specimen ZPAL T.25 KOW R061, from Kowala Quarry, Frasnian						
Mean	0.603	0.450	0.115	–	–	0.491
Standard deviation	0.0907	0.0615	0.0295	–	–	0.1402
N	31	31	31	8	3	31
Minimal value	0.40	0.34	0.06	0.10	0.46	0.16
Maximal value	0.74	0.54	0.16	0.18	0.74	0.84
V	0.151	0.137	0.256	–	–	0.286
Specimen ZPAL T.25 GC007–010, Góra Cmentarna, Early Frasnian						
Mean	0.623	0.410	0.139	–	–	0.385
Standard deviation	0.1214	0.0723	0.0397	–	–	0.1274
N	33	33	33	3	–	33
Minimal value	0.4	0.26	0.10	0.12	–	0.20
Maximal value	0.9	0.54	0.16	0.14	–	0.80
V	0.195	0.176	0.286	–	–	0.331

All measurements in millimetres. N, number of measurements; *V*, coefficient of variation; Max LD, longest lumen diameter; Min LD, shortest lumen diameter; DWT, double wall thickness; PD, pore diameter; PS, pore spacing; TS, tabulae spacing.

Material. Jaźwica (sets J and K): three coralla, six thin sections (ZPAL T.25 JAZ 020, JAZ 025.1–3; J032–033); Kowala (set A): six coralla from biostromal complex, 18 thin sections, 33 acetate peels (ZPAL T.25 AX. 001.1–2, KOW 4.005, KOW 6.001, KOW 12.001.1–3; K211–213, K293–300); Laskowa: three poorly preserved coralla, five thin sections (ZPAL T.25 L030–031, L033–034, L090).

Description. A laminar-irregular coralla, thickening in the middle, the largest one measures $35 \times 20 \times ?$ mm, composed of long, meandering corallites. They are long, strongly curved, pointing at acute angle to the surface. In cross section, they are kidney-shaped, elongated, oval, crescentic, rarely irregularly polygonal. Walls thin, transversely even, longitudinally uneven; median line invisible. Tabulae flat or convex, rarely slightly curved, spaced regularly and showing weakly growth periodicity (see growth periodicity chapter). The neighbouring corallites show different growth pattern (for details see chapter on growth periodicity). Mural pores round, rarely slightly oval, uniserial, on the shorter walls or in/near the corners. Pores spaced rather regularly, with the interval varying 0.48–0.70 mm, sporadically more. Their length varies from 0.08 to 0.12 mm. The distribution of septal apparatus is irregular, in some calyces absent or weakly developed and in others up to 8–10 spines present on one cross section of corallite; rarely only a *Hauptdorn* is present. Septal spines often blunt, clavate, rarely sharp-edged. The juvenile corallites are thin-walled, without septal apparatus, with tabulae spaced distantly.

Measurements. Table 2.

Remarks. The type material was never established, and the status of the original collection of Milne-Edwards and Haime is unknown to the present author. This species seems to be very similar to *A. suborbicularis*, and after the investigation of the type material, it may happen that the name of *A. compressus* will be considered as junior subjective synonymy of *A. suborbicularis*. There are two faint differences between these species: one is the arrangement of corallites (Lecompte 1933, p. 28), which is weakly seen on described here material and intracolonial variation, much lower than in *A. suborbicularis*. The above-described specimens show no significant differences with the material described by Lecompte (1933), except for the shape of corallite lumina, which in the material from the Holy Cross Mts., is more variable (e.g. rounded lumina occur more frequently).

The intracolonial variation of lumen diameter is relatively low (comparing to other species of the genus). In contrast, very high variation coefficient of tabulae spacing in corallum ZPAL T.25 KOW 12.001 is caused by measurements that were taken also in the juvenile parts of corallites, where tabulae are spaced distantly. The fluctuations of the double wall thickness can be a result of measuring on acetate peels; therefore, it must be treated not as surely as measurements on a thin section. The growth pattern of this species was described by Zapalski (2007*b*) and Zapalski *et al.* (2012).

Occurrence. Kostomłoty Transitional Zone (Givetian), Southern Kielce Subregion, Devonshire, Ardennes (Frasnian).

Alveolites edwardsi frasnianus Nowiński, 1992

```
  * v. 1992 Alveolites edwardsi frasnianus sp. n. Nowiński,
              pp. 192, 204, fig. 6A–B.
    v. 2003 Alveolites edwardsi frasnianus Nowiński;
              Nowiński, p. 141, pl. 83, fig. 1.
```

Type material. Paratypes ZPAL T.25: J037–038, J075–076, Institute of Paleobiology PAS, Warsaw.

TABLE 2. Intracolonial variation in *Alveolites compressus* Milne-Edwards and Haime.

	Max LD	Min LD	DWT	PD	PS	TS
Specimen ZPAL T.25 KOW AX.001, from Kowala Railroad Section, Frasnian						
Mean	0.732	0.536	0.155	0.161	0.622	0.403
Standard deviation	0.0957	0.0603	0.0527	0.0241	0.1222	0.1178
N	40	40	397	33	22	109
Minimal value	0.56	0.44	0.02	0.12	0.48	0.14
Maximal valuc	0.94	0.70	0.34	0.20	1.00	0.92
V	0.131	0.113	0.340	0.149	0.197	0.292
Specimen ZPAL T.25 KOW 12.001, from Kowala Railroad Section, Frasnian						
Mean	0.794	0.358	0.143	–	–	0.438
Standard deviation	0.1211	0.0564	0.0386	–	–	0.2211
N	37	37	32	9	–	32
Minimal value	0.5	0.22	0.08	0.10	–	0.22
Maximal value	1.04	0.56	0.2	0.14	–	1.26
V	0.153	0.158	0.271	–	–	0.505

Measurements in millimetres. For abbreviations, see Table 1.

Material. Jaźwica/Góra Łgawa: two coralla, four thin sections (ZPAL T.25: J037–038, J075–076).

Remarks. As the holotype (from the late Frasnian of Domaszowice) seems to be missing, the neotype should probably be established, but small fragments analysed here do not allow to analyse the biometry of this species statistically. The species was adequately described and illustrated by Nowiński (1992).

Occurrence. Kostomłoty Transitional Zone, Southern Kielce Subregion (Late Frasnian).

Alveolites maillieuxi Salée *in* Lecompte, 1933
Figure 9A–F

    * v. 1933 *Alveolites maillieuxi* Salée; Lecompte, p. 36, pl. 3, figs 1–2.
  v non 1933 *Alveolites maillieuxi* var. *cavernosa*; Lecompte, p. 38, pl. 3, fig. 4.
    ? 1952 *Alveolites maillieuxi* Salée; Sokolov, p. 93, pl. 24, fig. 3.
  non 1955 *Alveolites maillieuxi* Salée; Kraevskaya, p. 200, pl. 28, fig. 3.
  non 1956 *Alveolites maillieuxi* Salée; Dubatolov, p. 102, pl. 4, fig. 6a–b.
    1958 *Alveolites maillieuxi* Salée; Stasińska, p. 210, pl. 22, figs 1–2.
    1959 *Alveolites maillieuxi* Salée; Dubatolov, p. 146, pl. 17, fig. 2.
    ? 1975 *Alveolites maillieuxi* Salée; Khaiznikova, p. 64, pl. 11, fig. 1.
    1975 *A. maillieuxi*; Brice, Bigey, Mistiaen, Poncet and Rohart, p. 145.
    1976 *Alveolites maillieuxi* Salée; Stasińska and Nowiński, p. 303, pl. 22, figs 1–2.
  v p 2003 *Alveolites maillieuxi* Salée; Nowiński, p. 142, pl. 85, figs 1–3.
  . 2007 Alveolites maillieuxi Salée; Hubert, Zapalski, Nicollin, Mistiaen and Brice, p. 250.

Type material. MRHN 7069, thin section a322, from Han s. Lesse, Givetian, Gib, Belgium (Lecompte 1933, pl. 3, fig. 3, non 3a) and paralectotypes, Bruxelles, Belgium.

Material. Czarnów (set C): one corallum, six thin sections (ZPAL T.25 C001–006); Jaźwica (probably sets I–K): four coralla, seven thin sections (ZPAL T.25: JAZ 018.1–2, ZPAL T.25 JAZ 024, J014–J015, J065–067); Kowala (set A and rubble): five coralla, 13 thin sections (ZPAL T.25: KOW 9.003.1–2, KOW 9.010.1–3, KOW 9.010, KOW 10.001, KOW R. 042.1–3, KOW 043; KO E6.1–2), and three thin sections (ZPAL T.25 STA KL 001–003).

Description of the lectotype. Coralla massive. Two types of corallites occur within the corallum; they are grouped in aggrega-

tions: (1) thin-walled, polygonal in cross section (favositid-like) with polygonal lumina and (2) thick-walled, alveolitid, semicircular, with strongly rounded lumen. Also, two types of walls occur: thin, without median line visible between the corallites of the first type and thick, with crescentic shaped, discontinuous, strongly uneven median line, composed of large granulae of dark calcite. Mural pores frequent, distributed and spaced irregularly, their diameter varies from 0.14 to 0.20 mm, spacing varies 0.6–0.8 mm, and they are sometimes closed by poral plate. Tabulae numerous, usually flat, slightly concave or convex, weakly inclined or folded, sometimes incomplete, distributed rather regularly. Septal apparatus present, but spines are distributed irregularly – in some corallites few, in some absent (in the zone of polygonal corallites spines are not observed). They are short, sharp, conical or blunt, button-like, up to 3–4 visible on one cross section of corallite. In several corallites, a single, strong spine (*Hauptdorn*) is present on the longer wall.

Measurements. Table 3.

Description of the material from the Holy Cross Mts. Coralla discoidal, 55 mm of the largest diameter. Corallites straight or slightly bent, in certain zones of corallum polygonal and thin-walled and in others elongated, lens-shaped, shield-shaped or irregular. Walls thin, variable, uneven. Median line rarely visible. Round mural pores rather frequent (in specimens from Kowala), or rare (in specimens from Jaźwica); distributed irregularly. Tabulae numerous, usually flat, slightly concave or convex, weakly inclined or bent, sometimes incomplete, distributed rather regularly. Septal apparatus weakly developed (in specimens from Kowala), or very well developed (in specimens from Jaźwica); septal spines short, wide at the base, conical, sharp or blunt; distributed irregularly, often absent on cross section of corallite, in some specimens arranged in rows.

Measurements. Table 4.

Discussion. The material from the Holy Cross Mountains shows remarkable similarities to the type specimen of Lecompte. The distinctness is in more visible differentiation of corallites in the type material and longer and sharper spines. In the type specimen, mural pores are similarly frequent as in the material from Kowala, but the thin section of Lecompte did not allow making precise measurements; therefore, they lack in Table 3, similarly as in the material from Jaźwica, where pores occur, but their measurements are impossible. The intracolonial variation of this species is similar to the type species of the genus, and only corallite lumina are fairly higher variable.

Occurrence. Kostomłoty Transitional Zone (Late Givetian–Early Frasnian), Southern Kielce Subregion (Early Frasnian), Łysogóry Region (Eifelian–Givetian), Radom–Lublin Region (Givetian), Boulonnais (Givetian), Ardennes (Givetian), Kuznetsk Basin (Givetian).

TABLE 3. Intracolonial variation in *Alveolites maillieuxi* Lecompte.

	Max LD	Min LD	DWT type 1	DWT type 2	PD	TS
Specimen MRHN 7069, thin section a322, Han s. Lesse, Givetian, Gib, Belgium; lectotype						
Mean	0.602	0.416	0.043	0.148	–	0.406
Standard deviation	0.087	0.082	0.0156	0.049	–	0.084
N	35	35	31	32	7	38
Minimal value	0.42	0.28	0.02	0.06	0.14	0.30
Maximal value	0.8	0.6	0.08	0.26	0.22	0.64
V	0.144	0.197	0.363	0.331	–	0.207

Measurements in millimetres. For abbreviations, see Table 1.

TABLE 4. Intracolonial variation in *Alveolites maillieuxi* Lecompte.

	Max LD	Min LD	DWT	PD	PS	TS
Specimen ZPAL T.25 KOW 9.003, from Kowala, Frasnian						
Mean	0.681	0.434	0.075	0.165	0.701	0.525
Standard deviation	0.1090	0.0699	0.0236	0.0200	0.1348	0.1515
N	33	33	33	31	31	31
Minimal value	0.44	0.34	0.04	0.14	0.48	0.30
Maximal value	0.94	0.60	0.14	0.20	1.1	0.84
V	0.160	0.161	0.316	0.121	0.192	0.289
Specimen ZPAL T.25 J065–067, from Jaźwica, Frasnian						
Mean	0.616	0.435	0.104	–	–	0.644
Standard deviation	0.1208	0.0831	0.0252	–	–	0.1533
N	33	33	33	–	–	33
Minimal value	0.38	0.26	0.06	–	–	0.40
Maximal value	0.82	0.66	0.18	–	–	0.90
V	0.196	0.191	0.243	–	–	0.289

Measurements in millimetres. For abbreviations, see Table 1.

Alveolites multispinosus Dubatolov, 1959
Figure 10A–D

*. 1959 *Alveolites multispinosus* sp. nov. Dubatolov,
p. 143, pl. 48, fig. 1.
? 1969b *Alveolites multispinosus* Dubatolov; Stasińska,
p. 771.

Type material. Specimen 1/23 (holotype) and three paratypes, coll. V. N. Dubatolov, (?) All-Russia Petroleum Research Exploration Institute, St. Petersburg, Russia (Dubatolov 1959, pl. 48, fig. 1).

Material. Kowala Railroad Section: one incomplete corallum from biohermal complex, two thin sections (ZPAL T.25 KOW 4.004.1–2).

Description. Corallum massive, irregular, 55 mm of the largest diameter. Corallites long, irregularly twisted, meandering. Coral-lites in cross section subcircular, oval, kidney-shaped, rarely shield-like or irregular. Walls thin, rather even, with discontinuous median line visible in some places. On the cross sections, median line is usually visible as dark, crescent shape, on the longitudinal sections as thin, dark line. Connecting pores very rare, having at least 0.10–0.14 mm in diameter. Tabulae numerous, complete, convex or flat, rarely folded, spaced rather regularly. Septal apparatus very well developed, spines are numerous (up to seven on one cross section of a corallite), often long and sharp, rarely blunt.

Measurements. Table 5.

Discussion. A. *multispinosus* has originally been described from the Givetian of Kuznetsk Basin. The corallum described above strongly resembles the type material of Dubatolov (1959), with the only comment that the pore

FIG. 9. A–B, *Alveolites maillieuxi* Salée *in* Lecompte, 1933, lectotype. Han-sur-Lesse, Ardennes, Belgium; Givetian. Specimen MRHN 7069. A, transverse section. B, longitudinal section. C–D, *Alveolites maillieuxi* Salée *in* Lecompte, 1933. Jaźwica Quarry, rubble, Holy Cross Mountains, Poland; Frasnian. Specimen ZPAL T.25 J065–067. C, transverse section, D, longitudinal section. E–F, *Alveolites maillieuxi* Salée *in* Lecompte, 1933. Kowala Railroad Section, Holy Cross Mountains, Poland; rubble of sets A–C, Early Frasnian. Specimen ZPAL T.25 KOW R.042. E, transverse section. F, longitudinal section. Scale bars represent 1 mm.

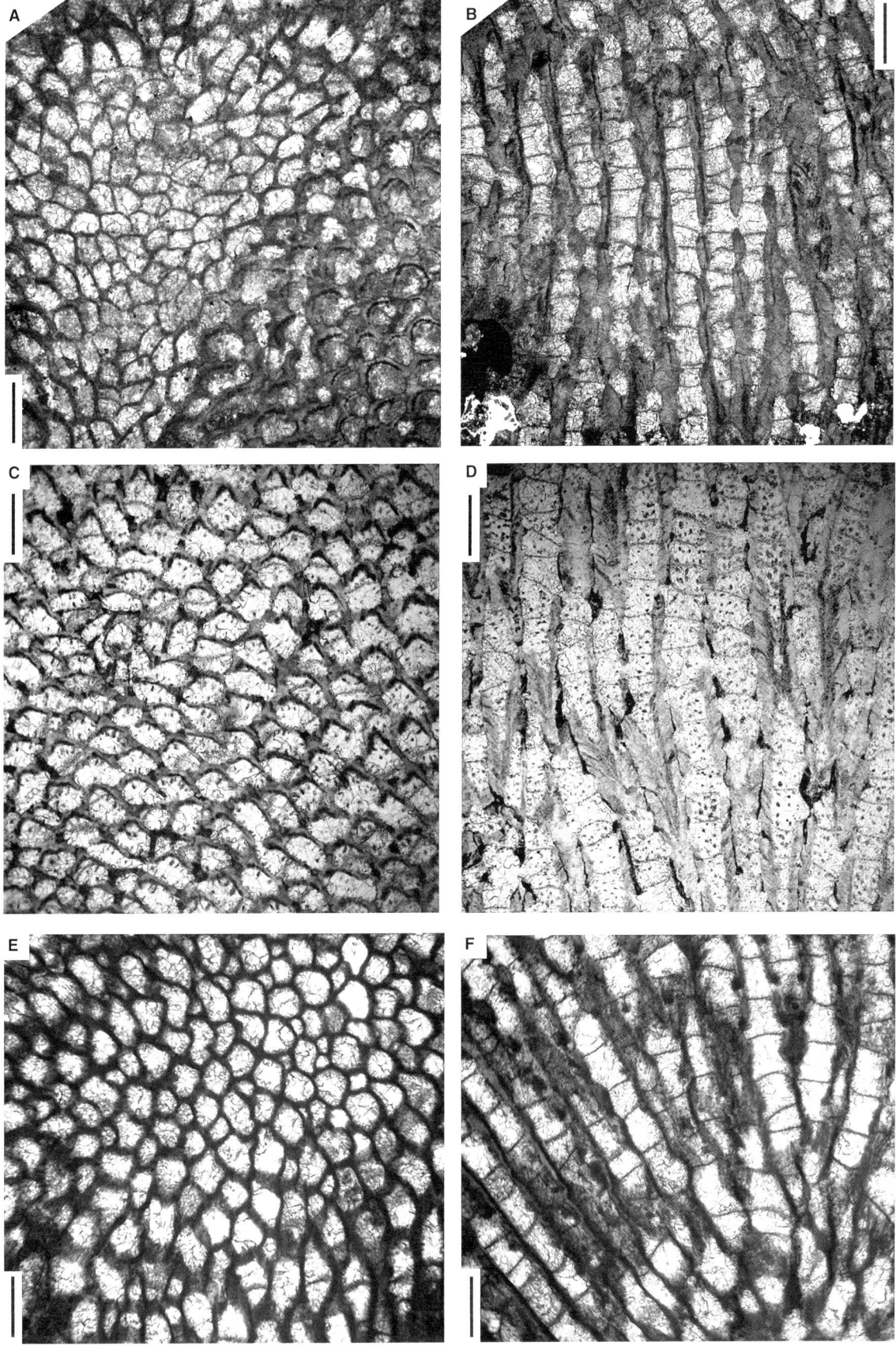

TABLE 5. Intracolonial variation in *Alveolites multispinosus* Dubatolov.

	Max LD	Min LD	DWT	PD	PS	TS
Specimen ZPAL T.25 KOW 4.004, from Kowala, Frasnian						
Mean	0.584	0.481	0.153	–	–	0.419
Standard deviations	0.0933	0.0542	0.0308	–	–	0.0784
N	30	30	30	2	–	30
Minimal value	0.46	0.36	0.08	0.10	–	0.28
Maximal value	0.86	0.60	0.22	0.14	–	0.60
V	0.196	0.191	0.243	–	–	0.289

Measurements in millimetres. For abbreviations, see Table 1.

spacing in the specimen from the Holy Cross Mountains is unknown.

The investigated corallum may contain parasites belonging to the genus *Chaetosalpinx*, but only one section was obtained (Fig. 10A) and determination is doubtful.

The intracolonial variation in the *A. multispinosus* is more balanced than in other species of the genus, that is *V* values of lumen diameters are higher than average, while *V* values of tabulae spacing and double wall thickness are lower than average.

Occurrence. Southern Subregion of the Kielce Region (Frasnian) and Western Pomerania (Givetian), Poland; Kuznetsk (Givetian), Russia.

Alveolites parvus Lecompte, 1939
Figure 11A–F

* v. 1939 *Alveolites parvus* sp. n. Lecompte, p. 43, pl. 6, figs 1–3.
non 1952 *Alveolites parvus* Lecompte; Sokolov, p. 95, pl. 25, figs 1–2.
v ? 1953 *Alveolites parvus* Lecompte; Stasińska, p. 230, pl. 3, fig. 2.
? 1979 *Alveolites parvus* Lecompte; Hladil, p. 180, fig. 3.
v. 2003 *Alveolites parvus* Lecompte; Nowiński, p. 143, pl. 86, fig. 3–4.
? 2003 *Alveolites parvus* Lecompte; Fernández-Martínez and Mistiaen, p. 262, pl. 17, figs 1–5, pl. 18, figs 1–4.
non 2005 *Alveolites parvus* Lecompte; Tsyganko and Lukin, p. 22, pl. 1, figs 6–7.
. 2007 *Alveolites parvus* Lecompte; Hubert, Zapalski, Nicollin, Mistiaen and Brice, p. 250.

Type material. Specimen MRHN 2 (lectotype), specimen MRHN 624 (paralectotype), thin section a400 (lectotype) and paralectotypes, Olloy, Belgium; Frasnian (lectotype: Lecompte 1939, pl. VI, fig. 2).

Material. Jaźwica (set J or K): one corallum, two thin sections (ZPAL T.25 JAZ 010.1–2); Laskowa: three coralla, 11 thin sections (ZPAL T.25: L003–007, L015–017, L095–097); Psie Górki: two coralla, two thin sections (ZPAL T.25 STA PG 007–008); Sowie Górki: two coralla, four thin sections (ZPAL T.25 SG 100–103); Szczukowskie Górki: two coralla, two thin sections (ZPAL T.25 STA SWG 001–002); Wietrznia: four coralla, five thin sections (ZPAL T.25 STA W 004–008).

Description of the type material. Coralla small, massive. Corallites short and medium, curved, in cross section shield-like and alveolitoid, subpentagonal, subrectangular, subtriangular. They reach the surface of the corallum at acute angle. Lumina with rounded angles. Walls moderately thin, strongly uneven, median line may be visible as thin, dark, discontinuous line, often feebly granular. Pores scarce, usually closed by poral plate. Tabulae numerous, flat, concave, rarely slightly inclined, spaced evenly. Septal apparatus very well developed, as long, thin, sharp spines on all walls; often main, strong spine (*Hauptdorn*) occurs. Septal spines horizontal, sometimes bent upwards or downwards. The length of septal spines may reach half of the minimal lumen width.

Measurements. Table 6.

Description of the material from the Holy Cross Mts. Coralla small, massive, irregular, up to about 50 mm of the largest diameter. Corallites long, twisted; subtriangular, oval, alveolitoid, feebly shield-like in cross section. In the initial stages, they are larger, subrectangular or polygonal in cross section, with thinner walls and lacking septal spines. Corallites reach the surface of coralla at various, most often acute angle. Walls moderately thick, thickened at junctions of walls, in places uneven, with sudden thickening supporting several small spines. Median line visible as very thick, crescentic, discontinuous shape on upper lips of corallites. Connecting pores very scarce, their diameter probably varies between 0.10 and 0.20 mm, and they are often closed by poral plate; their spacing unknown. Tabulae thin, flat and feebly concave, usually complete, spaced rather regularly, but in the corallites placed in peripheries of coralla, their spacing is larger. Septal spines numerous, small, sharp, *hauptdorn* often present.

Measurements. Table 7.

Discussion. The material from the Holy Cross Mts. differs from the Belgian specimens by having thicker walls and less developed septal apparatus; in contrast, *Hauptdorn*

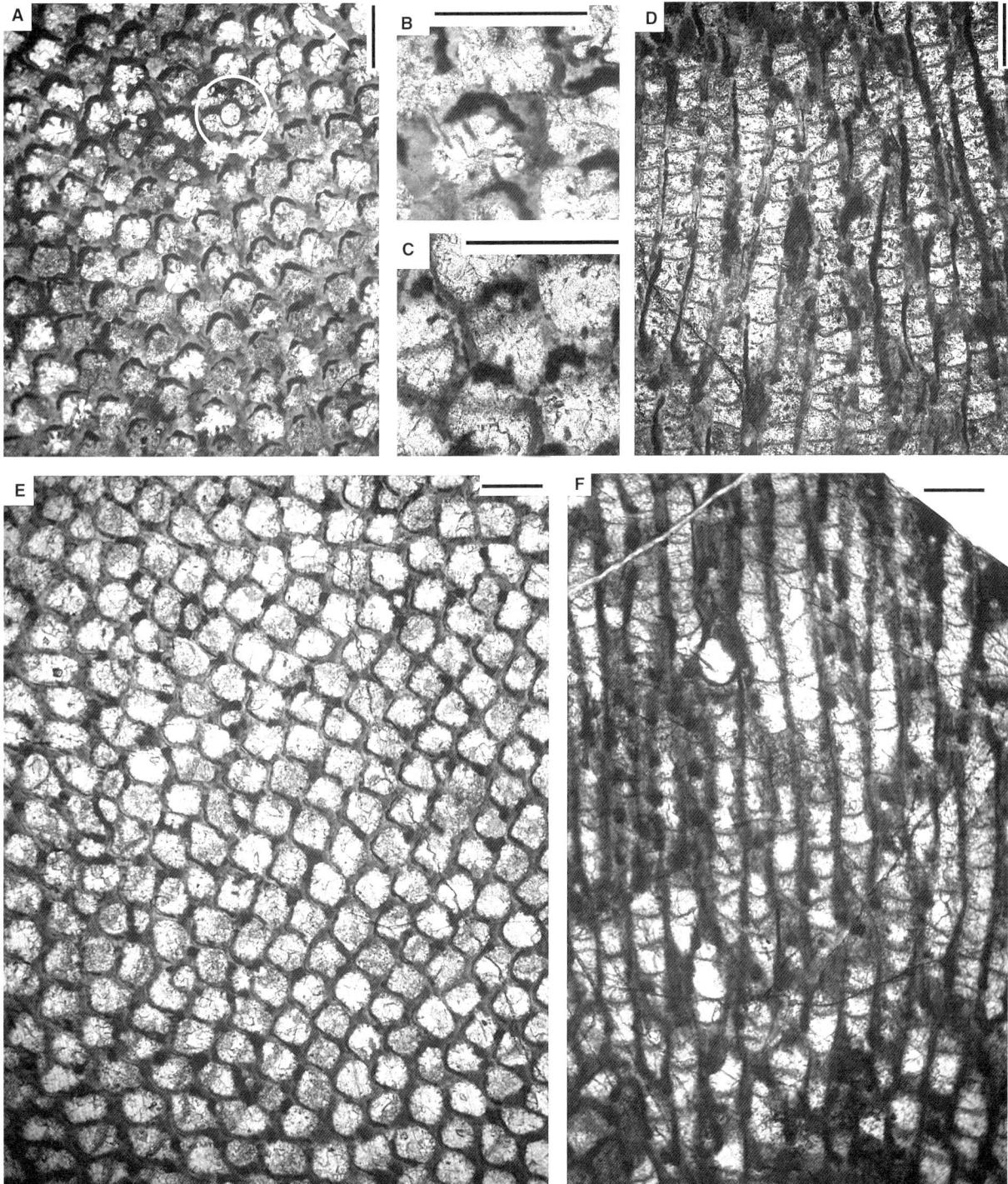

FIG. 10. A–D, *Alveolites multispinosus* Dubatolov, 1959. Kowala Railroad Section, Holy Cross Mountains, Poland; set A, Early Frasnian. Specimen ZPAL T.25 KOW 4.004. A–C transverse section (possible *Chaetosalpinx* cf. *plusquelleci* sp. nov. in the circle). D, longitudinal section. B and C show variation of corallite shape and development of spines. E–F, *Alveolites regularis* Sokolov, 1952. Kowala Railroad Section, Holy Cross Mountains, Poland; set A, Early Frasnian. Specimen ZPAL T.25 KOW 11.002. E, transverse section. F, longitudinal section. Scale bars represent 1 mm.

occurs more frequently and appear to be more developed than in the type material. On the biometrical values given above, the larger corallites are visible, which, in the pres- ent author's opinion, is an effect of sectioning of the specimen not precisely perpendicularly to the axes of growth of corallites.

TABLE 6. Intracolonial variation in *Alveolites parvus* Lecompte.

	Max LD	Min LD	DWT	PD	PS	TS
Specimen MRHN 2, thin section a400, from Olloy, Frasnian; lectotype						
Mean	0.495	0.288	0.095	–	–	–
Standard deviation	0.0720	0.0376	0.0197	–	–	–
N	32	32	32	–	–	–
Minimal value	0.36	0.24	0.06	–	–	–
Maximal value	0.68	0.38	0.14	–	–	–
V	0.145	0.131	0.207	–	–	–
Specimen MRHN 624, thin section a401, from Olloy, Frasnian; paralectotype						
Mean	0.458	0.303	0.100	–	–	0.330
Standard deviation	0.0579	0.0341	0.0246	–	–	0.0941
N	50	50	50	8	–	38
Minimal value	0.50	0.24	0.08	0.06	–	0.22
Maximal value	0.58	0.40	0.16	0.14	–	0.60
V	0.126	0.113	0.246	–	–	0.286

Measurements in millimetres. For abbreviations, see Table 1.

TABLE 7. Intracolonial variation in *Alveolites parvus* Lecompte.

	Max LD	Min LD	DWT	PD	PS	TS
Specimen ZPAL T.25 SG100–103, from Sowie Górki, Frasnian						
Mean	0.488	0.310	0.135	–	–	0.282
Standard deviation	0.0814	0.0492	0.0318	–	–	0.1258
N	31	31	31	–	–	31
Minimal value	0.30	0.22	0.08	–	–	0.12
Maximal value	0.70	0.44	0.20	–	–	0.70
V	0.167	0.159	0.236	–	–	0.446
Specimen ZPAL T.25 L015–017, from Laskowa Góra Quarry, Givetian						
Mean	0.579	0.331	0.179	–	–	0.354
Standard deviation	0.0922	0.0344	0.0350	–	–	0.0987
N	35	35	35	–	–	35
Minimal value	0.4	0.26	0.12	–	–	0.08
Maximal value	0.74	0.4	0.26	–	–	0.5
V	0.159	0.104	0.196	–	–	0.279

Measurements in millimetres. For abbreviations, see Table 1.

The intracolonial variation of the discussed species is variable, low for lumen diameters, very high for tabulae spacing. In contrast, the intraspecific variation is rather high, both in quantitative and in qualitative features. For example in the type specimens, corallum MRHN a401 has shield-like corallites dominating, while in the lectotype, MRHN a400, lens-shaped and alveolitoid corallites dominate. Another example of high intraspecific variation can be differences in V values between the type material and specimen ZPAL T.25 SG100–103.

Remarks. Sokolov (1952) established the lectotype of this species on the basis of illustrations of Lecompte (1933). Such an act, without investigation of the material, seems to be highly undesirable, because of very poor preservation of chosen lectotype (as it is in this particular case, but also in the case of 'Alveolites' obtortus). Moreover, there exists only transverse section of the lectotype and several important taxonomical features (such as pore diameter, pore spacing, tabulae spacing and their form) cannot be recognized on the lectotype. Therefore, another

FIG. 11. A–B, *Alveolites parvus* Lecompte, 1939, paralectotype. Olloy B9, Frasnian (F2g) Belgium. Specimen MRHN 624, thin sections a401. A, transverse section (circles mark shield-like corallite). B, longitudinal section. C–D, *Alveolites parvus* Lecompte, 1939. Laskowa Góra Quarry, Holy Cross Mountains, Poland; Laskowa Beds (?), Givetian. Specimen ZPAL T.25 L003–007. C, transverse section; D, longitudinal section. E–F, *Alveolites parvus* Lecompte, 1939. Jaźwica Quarry, Holy Cross Mountains, Poland; set K (?), Early Frasnian. Specimen ZPAL T.25 JAZ 007. E, transverse section, F, longitudinal section. Scale bars represent 1 mm.

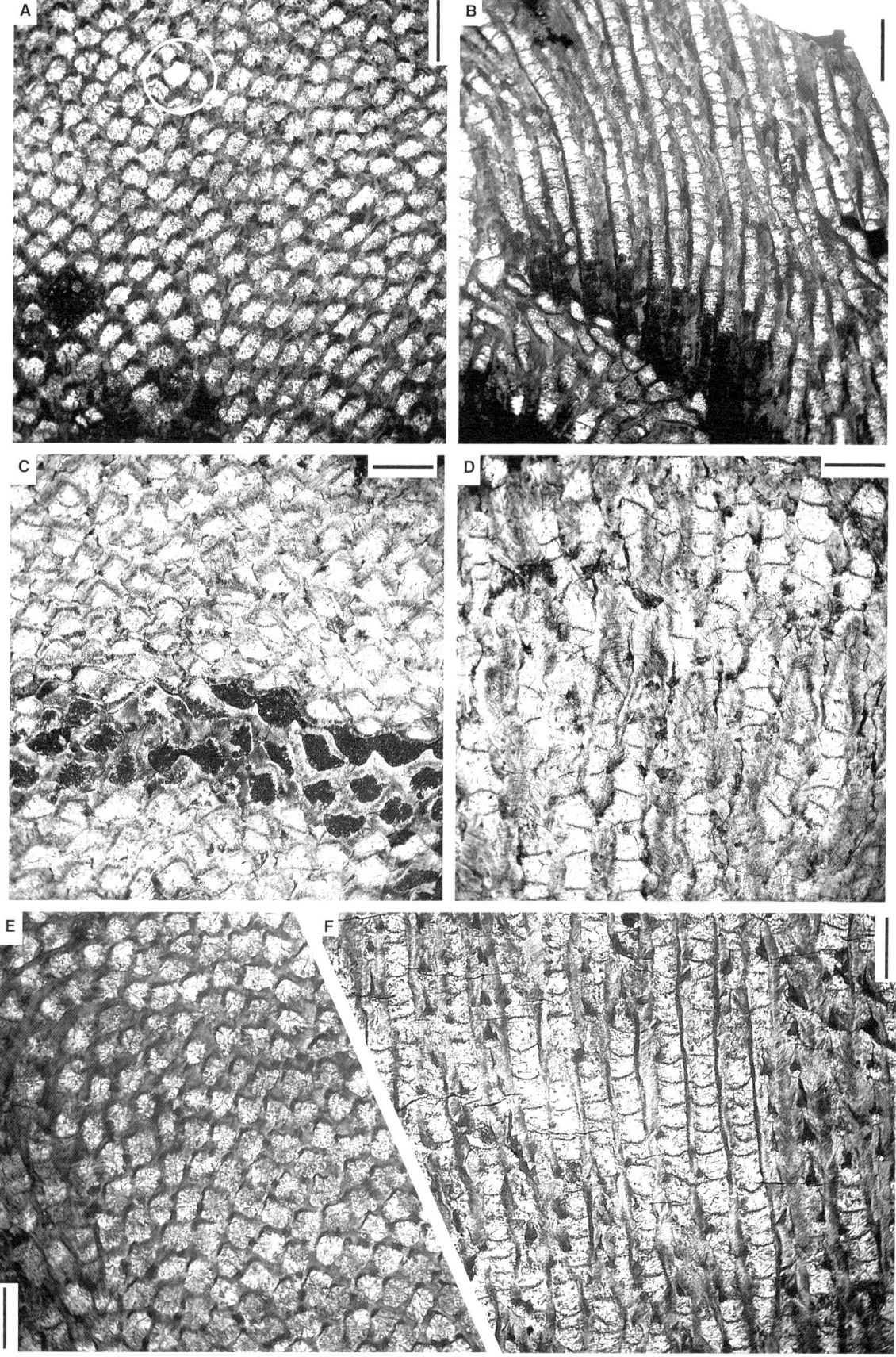

specimen (one of paralectotypes), incomparably much better preserved, with correct transverse and longitudinal sections have been analysed.

Occurrence. Kostomłoty Transitional Zone (Givetian), Kielce Region (Frasnian), Ardennes, Moravia (?Frasnian), Iran (?Frasnian).

Alveolites regularis Sokolov, 1952
Figure 10E–F

```
  *. 1952   Alveolites regularis sp. n. Sokolov, p. 103, pl. 27,
                figs 3, 4.
    1981a   Alveolites regularis Sokolov; Hladil, p. 29, pl. 1,
                figs 3–5.
  v. 1992   Alveolites regularis Sokolov; Nowiński, p. 194,
                text-fig. 7.
  v. 2003   Alveolites regularis Sokolov; Nowiński, p. 144,
                pl. 89, fig. 1.
```

Type material. Probably in the collections of All-Russia Petroleum Research Exploration Institute (coll. Domrachev), St. Petersburg, Russia, repository and specimen number not indicated in the original paper (Sokolov, 1952, pl. 27, figs 3–4).

Material. Kadzielnia (set A): one corallum, one thin section (ZPAL T.25 STA KA 001); Kowala (set A and rubble): six coralla, 15 thin sections, two acetate peels (ZPAL T.25: KOW 9.A03.01, KOW 11.002, KOW R.040.1–3, KOW R.075.1–5, K113–115, K410–411); Laskowa: one corallum, four thin sections (ZPAL T.25 L065–068).

Description. Corallum large, massive, bulbous, measuring about $100 \times 60 \times 50$ mm, with irregular surface. Calyces rhomboidal or rectangular, with rounded edges. Corallites most often rhomboidal in cross section, in some parts of corallum rectangular to lens-shaped (alveolitid) and rarely irregular. The size of corallite lumen (in cross section) varies from 0.58×0.62 to 0.80×0.92 mm, rarely reaching 1.00 mm. In longitudinal section, corallites straight in some parts, usually meandering. Walls thin and moderately thick, with double wall thickness variation from 0.04 to 0.14 mm (most commonly 0.06–0.08 mm). Walls slightly uneven, median line sometimes visible; on the longitudinal section as thin, dark line, on the cross section as dark, crescentic shape on the upper lip of corallite. Tabulae thin, spaced irregularly, their mean spacing is 0.43 mm, and ranging from 0.22 to 0.72 mm (see the chapter on growth periodicity). They are flat, curved, inclined, concave or convex, often incomplete. Connecting pores rare, medium and large, round, 0.14–0.22 mm in diameter. Septal spines very rare, irregularly distributed, small, button-like or long and sharp.

Remarks. As having technical difficulties in sectioning the material (not straight corallites, longitudinal sections on one specimens, cross sections of the other), the study of intracolonial/intraspecific variation of this species was not undertaken; however, it can be stated that variation in size and shape of corallites is rather low. The variation coefficient of tabulae spacing for corallum KOW R.075 is 0.277 and therefore similar to other alveolitids. The material described above seems to be entirely conspecific with the species described by Sokolov (1952). All biometrical characters (corallite size, wall thickness, pore diameter and tabulae spacing) appear to be the same. The growth pattern of this species was described by Zapalski *et al.* (2012).

Occurrence. Kostomłoty Transitional Zone (Givetian), Northern and Southern Kielce Regions (Early Frasnian), Western slopes of Ural Mts. (Late Frasnian), Moravia (Frasnian).

Alveolites? *obtortiformis* sp. nov.
Figure 12A–D

```
  * v p 1939   Alveolites obtortus sp. n. Lecompte, p. 42, pl. 6,
                  fig. 4–7.
    non 1953   Alveolites obtortus Lecompte; Stasińska, p. 231,
                  pl. 3, fig. 3.
      ? 1959   Alveolites obtortus Lecompte; Dubatolov
                  pp. 144–145; pl. 48, figs 3A–B, 4A–B.
     vp 1992   Alveolites obtortus Lecompte; Nowiński,
                  p. 194.
     vp 2003   Alveolites obtortus Lecompte; Nowiński, p. 142,
                  non pl. 86, figs 1–2.
      p 2007   Alveolites obtortus Lecompte; Hubert, Zapalski,
                  Nicollin, Mistiaen and Brice, p. 250.
```

Derivation of the name. Because the new species is erected on the basis of part of the original material of *Alveolites obtortus* Lecompte.

Type strata. Fm. des Grands Breux, Frasnian.

Type locality. Senzeille, Ardennes, Belgium.

Diagnosis. Coralla massive, corallites irregularly prismatic and elongated in cross section. Corallite lumina 0.495 (± 0.07) $\times$ 0.67 (± 0.07) mm in diameter. Walls even, 0.13 (± 0.03) mm thick. Tabulae flat, usually complete, spaced unevenly. Round connecting pores in corners, 0.14–0.18 mm in diameter. Septal spines absent.

FIG. 12. A–B, *Alveolites?* *obtortiformis* sp. nov., holotype. Senzeille, Belgium; Fm. des Grands Breux Frasnian, F2h (local stratigraphy), specimen MRHN 1806, thin section a395. A, transverse section, B, longitudinal section. C–D, *Alveolites?* *obtortiformis* sp. nov. Miedzianka, Holy Cross Mountains, Poland; Frasnian. Specimen ZPAL T.25 MD 005–007. C, transverse section (circle marks possible *Chaetosalpinx*); D, longitudinal section. E–F, *Alveolites?* sp. Laskowa Góra Quarry, Holy Cross Mountains, Poland; Laskowa Beds, Givetian. Specimen ZPAL T.25 LAS 100. E, transverse section; F, longitudinal section. Scale bars represent 1 mm.

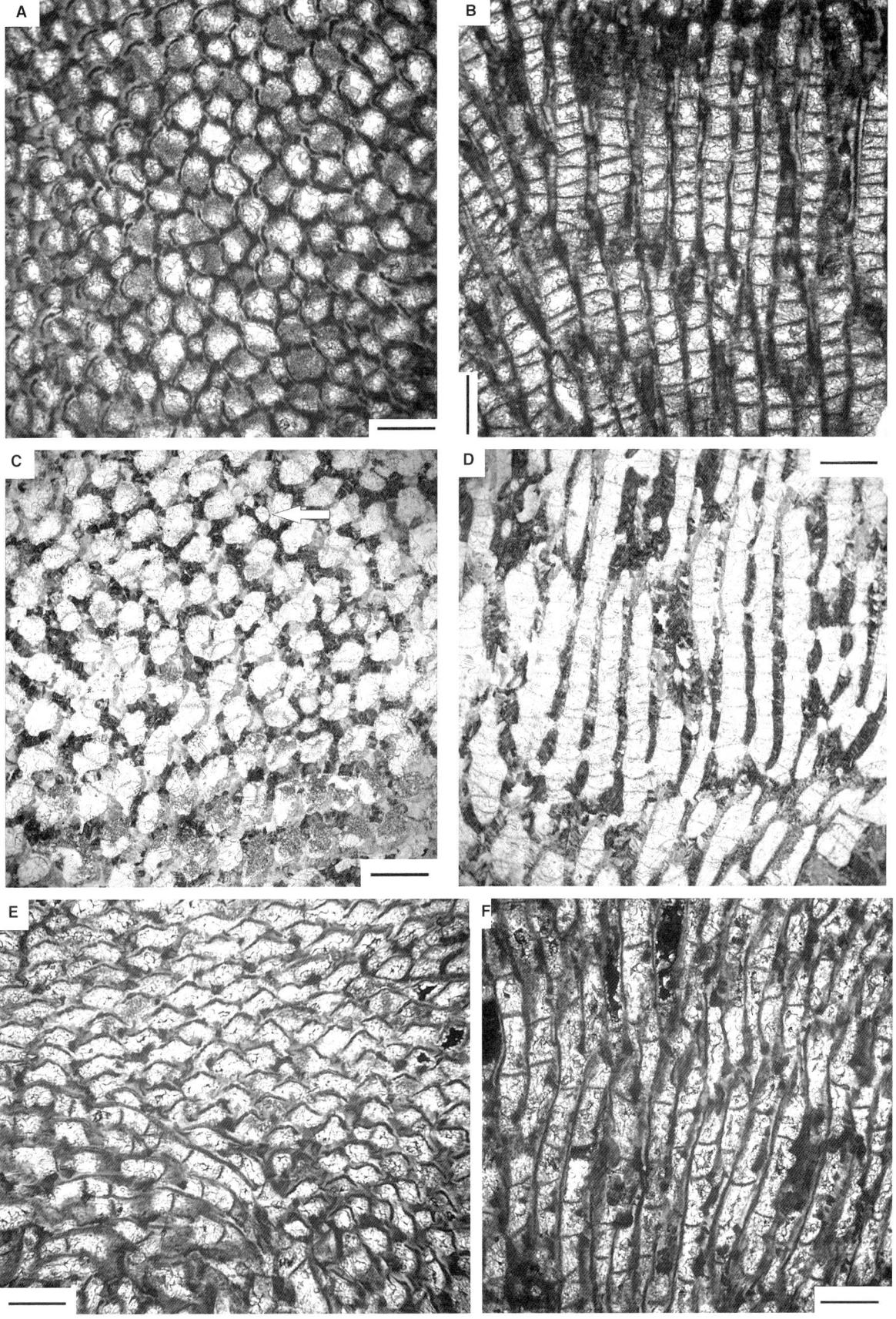

TABLE 8. Intracolonial variation in *Alveolites*? *obtortiformis* sp. nov.

	Max LD	Min LD	DWT	PD	PS	TS
Specimen MRHN 1806, thin section a395, from Senzeille, Frasnian; paralectotype						
Mean	0.666	0.495	0.135	–	–	0.444
Standard deviation	0.0695	0.0680	0.0333	–	–	0.270
N	32	32	33	2	–	52
Minimal value	0.50	0.38	0.08	0.14	–	0.22
Maximal value	0.82	0.60	0.20	0.18	–	1.20
V	0.104	0.137	0.247	–	–	0.609

Measurements in millimetres. For abbreviations, see Table 1.

Holotype. Specimen MRHN 1806 (Fig. 12A–B) thin section a395, from Senzeille, Belgium; Frasnian, F2h (old, local stratigraphy), Fm. des Grands Breux (Lecompte 1939, pl. 4, figs 3, 4 = Fig. 12A–B).

Other material. Miedzianka: one corallum, three thin sections (ZPAL T.25 MD 005–007); Trzemoszna: one corallum, two thin sections (ZPAL T.25 TR026–027).

Description of the holotype. Coralla massive. Corallites long, irregularly meandering, variable in shape (in cross section): most often irregular, semicircular, lens-shaped, subtriangular or subtrapezoidal. Corallite lumina with slightly rounded angles. Walls thin, even, with rather constant thickness. Median line visible as thick, crescentic dark line between corallites, most often only on only one side of corallite; it is discontinuous. Septal apparatus absent. Connecting pores very few, round, 0.14–0.18 mm in diameter, spacing unknown. Tabulae numerous, thin, most often flat, complete; rarely concave or convex, sometimes incomplete.

Measurements. Table 8.

Description of the material from the Holy Cross Mts. Coralla massive. Corallites long, usually straight, variable in shape (in cross section): most often polygonal rounded, rarely alveolitid or subrectangular. Corallite lumina with slightly rounded angles. Walls thin, even, with quite constant thickness, but in few places in corallum, they are longitudinally uneven. In the parts of corallum where walls are thicker, the corallite lumina are strongly rounded. Median line invisible. Connecting pores very few, round or oval, spacing unknown. Tabulae numerous, thin, most often flat, complete, rarely incomplete; they sometimes may be feebly convex. Septal apparatus absent.

Measurements. Table 9.

Discussion. The new species has been assigned to the genus *Alveolites* on the basis of corner placement of pores. Generic assignment of this material is very difficult, as corner pores may occur in *Palaeofavositinae* as well. Weak elongation of some of the corallites suggests, however, affinities to Alveolitidae, and within this family, it seems to be most similar to some representatives of *Alveolites*. The new species differs from other species of the genus by frequent polygonal corallites (usually elongated in other species) and much larger corallites. Most of representatives of this genus are distinguished by the 'alveolitid' shape of the corallites. Similar species are *Alveolites giganteus* Sokolov, from which the new species differs by much smaller corallites (up to 2.5 mm of the largest diameter in *A. giganteus*) and *Alveolites nalivkini* Sokolov, from which the new species differs by much smaller connecting pores (0.3–0.4 mm in *A. nalivkini*).

The new species is erected on the basis of material of Lecompte, originally assigned to *Alveolites obtortus*. The status of this species is very complicated. First of all, Lecompte (1939, p. 43) stated that septal spines in *A. obtortus* are well developed. Dubatolov (1959, p. 144) established the lectotype (Lecompte 1939, pl. 6, fig. 4) on the

TABLE 9. Intracolonial variation in *Alveolites*? *obtortiformis* sp. nov.

	Max LD	Min LD	DWT	PD	PS	TS
Specimen ZPAL T.25 MD 005–007, Miedzianka, Early Frasnian						
Mean	0.732	0.422	0.174	–	–	0.336
Standard deviation	0.0904	0.0505	0.0407	–	–	0.0708
N	22	22	22	2	–	22
Minimal value	0.56	0.34	0.10	0.14	–	0.20
Maximal value	0.9	0.50	0.24	0.14	–	0.50
V	0.124	0.120	0.234	–	–	0.211

Measurements in millimetres. For abbreviations, see Table 1.

basis of the Lecompte's illustration, without investigation of the material. This material is the only thin section of this corallum, and it is too thick to measure biometrical characters of this corallum; characteristic biometrical features of the lectotype cannot be recognized. On the other hand, septal apparatus seems to be lacking in this specimen, which is in contrast to the original idea of Lecompte (1939). All subsequent authors (e.g. Nowiński 1992) were describing 'A. obtortus' – following the original idea of Lecompte – as having spines. As the lectotype does not permit to recognize its features, probably a formal demand to the International Commission on the Zoological Nomenclature should be submitted to abandon Dubatolov's lectotype and establish a new one. After reinvestigating the Belgian material, it became obvious that the material of Lecompte belongs to at least two taxa; part of the material is assigned to the new species. Only very small part of Nowiński's *A. obtortus* has been assigned to this species in the present work, others were assigned here to various species of *Crassialveolites* (see synonymies), and part of the material was indeterminable. The specimen from Miedzianka is infested by *Chaetosalpinx? plusquelleci* sp. nov.

Occurrence. Central Kielce Subregion (Late Givetian – Frasnian), Ardennes (Frasnian).

Alveolites? tenuissimus Salée in Lecompte, 1933
Figure 13A–C

     * v. 1933 *Alveolites tenuissimus* Salée, nov. sp.; Lecompte,
            p. 42, pl. 4, figs 1, 1a, 2.
         1933 *Alveolites lemniscus* sp. n. Smith, p. 140, pl. 2,
            fig. 8; pl. 3, figs 1–3.
       v. 1939 *Alveolites tenuissimus* Salée; Lecompte, p. 59,
            pl. 10, figs 1–8.
   non v 1939 *Alveolites tenuissimus* Salée var. *spinosus* var.
            nov. Lecompte, p. 61, pl. 10, fig. 11, 11a.
    ? v 1939 *Alveolites tenuissimus* Salée var. *crassus* var.
            nov. Lecompte, p. 61, pl. 10, fig. 9, 9a.

       v. 1953 *Alveolites tenuissimus* Salée; Stasińska, p. 218,
            text-fig. 5, pl. 1, fig. 1.
         1970 *Tetralites tenuissimus* (Salée); Mironova,
            p. 126.
         1975 *A. tenuissimus*; Brice, Bigey, Mistiaen, Poncet
            and Rohart p. 145.
         1985 *Alveolites tenuissimus* Lecompte; Birenheide,
            p. 81.
         2002 *Alveolites (Subalveolites) tenuissimus* Salée;
            Mistiaen, Becker, Brice, Dégardin, Derycke,
            Loones and Rohart, p. 78.
         2002 *Alveolites (Subalveolites) tenuissimus* Salée;
            Mistiaen, p. 86.
       v. 2003 *Alveolites tenuissimus* Salée; Nowiński, p. 146,
            pl. 89, figs 2–3, pl. 90, fig. 1.
     . 2007 *Alveolites tenuissimus* Salée; Hubert, Zapalski,
            Nicollin, Mistiaen and Brice, p. 250.

Type material. Specimen MRHN a314 (lectotype) and paralectotypes, Bruxelles, Belgium (see Mironova 1970).

Material. Jaźwica/Góra Łgawa (Late Frasnian sets): 10 coralla, 26 thin sections (ZPAL T.25: J002, J005–08, J012–013, J034–036, J039–041, J044–047, J049–054, J057–058, J064, JAZ X01); Psie Górki: One corallum, one thin section (ZPAL T.25 STA PG A01).

Paralectotype. Specimen MRHN 1212, thin section a325, from Han-sur-Lesse, Belgium; Frasnian, F2g (old, local stratigraphy); Lecompte 1939, pl. 4, fig. 1, 1a.

Description of the paralectotype. Small, massive corallum, composed of strongly compressed corallites. Corallites rectangular in cross section, strongly elongated, rather uniform in shape, they are arranged in slightly inclined rows. In very basal parts of corallum, small oval or slightly irregular corallites may occur. Walls even, thin, with thin median line visible only on the longitudinal sections. Connecting pores very small, round, very rare. Septal apparatus absent. Tabulae flat, their spacing cannot be observed on the type specimen.

Measurements. Table 10.

Description of the material from the Holy Cross Mts. Small and medium size, bulbous coralla, up to 50 mm in diameter and

TABLE 10. Intracolonial variation in *Alveolites? tenuissimus* Lecompte.

	Max LD	Min LD	DWT	PD	PS	TS
Specimen MRHN 1212, thin section a325, Frasnian from Han-sur-Lesse, Belgium, syntype						
Mean	0.548	0.116	0.054	–	–	–
Standard deviation	0.0692	0.0207	0.0178	–	–	–
N	36	36	36	3	–	–
Minimal value	0.40	0.10	0.02	0.06	–	–
Maximal value	0.72	0.18	0.08	0.06	–	–
V	0.126	0.178	0.329	–	–	–

Measurements in millimetres. For abbreviations, see Table 1.

TABLE 11. Intracolonial variation in *Alveolites*? *tenuissimus* Lecompte.

	Max LD	Min LD	DWT	PD	PS	TS
Specimen ZPAL T.25 J006–007, Jaźwica, Frasnian						
Mean	0.510	0.120	0.031	–	–	–
Standard deviation	0.0538	0.0178	0.0165	–	–	–
N	34	34	34	–	–	–
Minimal value	0.42	0.1	0.02	–	–	–
Maximal value	0.62	0.16	0.06	–	–	–
V	0.105	0.148	0.532	–	–	–

Measurements in millimetres. For abbreviations, see Table 1.

about 70 mm in height, often encrusting organic remains. They are composed of strongly compressed, irregularly meandering corallites. Corallites rectangular in cross section, strongly elongated, often slightly bent, and they are arranged in slightly inclined rows, however not that regular as in the type specimen. Walls even, thin, with thin median line visible vaguely only on the longitudinal sections. Connecting pores invisible. Septal apparatus absent. Tabulae flat or folded, sometimes concave, spaced 0.25–0.40 mm.

Measurements. Table 11.

Discussion. A.? *tenuissimus* is an easily recognizable species by its characteristic, subrectangular shape of corallites. The specimens from the Jaźwica have nearly the same dimensions as the lectotype, but the corallite wall is somewhat thinner than in the type material. This difference may be an effect of the angle of sectioning. On the Holy Cross Mts. material connecting pores are invisible; tabulae spacing is unmeasurable owing to the meandering corallites and problems with obtaining correct longitudinal section (even when a longitudinal section is obtained, it is impossible to get two tabulae to measure the space between them). The problem of correct sectioning occurred probably in the specimen described by Lecompte as *A. tenuissimus* var. *crassus*, which is probably cut at an incorrect angle. The corallites in this species are very chaotically arranged, often twisted and meandering, and therefore very difficult to study.

Mironova (1970) described a new genus *Tetralites* with the type species *A. tenuissimus* Salcé. The most important

feature of this genus was rectangular shape of corallites. Hill (1981) included *Tetralites* (with a question mark) into the synonymy of *Kitakamiia*. A? *tenuissimus* cannot belong to *Kitakamiia* because the diagnostic feature of the latter genus is the chevron-shaped corallites in cross section, which is not the case for the discussed species, and *Kiitakamia* has sometimes occurring septal comb.

There are doubts concerning the authorship of this taxon (E. Fernández-Martínez, pers. comm. 2008). Both Smith and Lecompte described the same species as new in 1933. The paper by Lecompte bears the 31 July 1933 on the cover, while the volume with Smith's paper has no publication date. As there are summaries of meetings, it can be assumed that it was printed after the 8th November 1933 (there is a summary of the meeting of 8th November). Therefore, Lecompte's name has priority.

Occurrence. Southern Kielce Subregion (?Middle–Late Frasnian), Boulonnais (lower Frasnian), Ardennes (Frasnian).

Alveolites? sp.
Figure 12E–F

Material. Laskowa: one corallum, three thin sections (LAS 100.1–3).

Description. Corallum discoidal aberrant, measuring about 60 mm of the largest diameter. Corallites long, flattened, inclined to the corallum surface, irregularly twisted, in several places showing rosebud structure similar to the typical of *Roseoporella*. Corallites in cross section elongated, alveolitid, bean-

FIG. 13. A, *Alveolites*? *tenuissimus* Salée *in* Lecompte, 1933. Han-sur-Lesse, Belgium; Frasnian, F2g (local stratigraphy). Specimen MRHN 1212, paralectotype, thin section a325, transverse section. B, *Alveolites*? *tenuissimus* Salée *in* Lecompte, 1933. Jaźwica Quarry, Holy Cross Mountains, Poland; set K (?), Early Frasnian. Specimen ZPAL T.25 J 005, transverse section. C, *Alveolites*? *tenuissimus* Salée *in* Lecompte, 1933. Jaźwica Quarry, Holy Cross Mountains, Poland; set K (?) Frasnian. Specimen ZPAL T.25 J 064, longitudinal section. D–E, *Crassialveolites cavernosus* Lecompte, 1933. Posłowice, Holy Cross Mountains, Poland; Givetian. Specimen ZPAL T.25 PW 029. D, transverse section; E, longitudinal section. F–G, *Crassialveolites cavernosus* Lecompte, 1933, lectotype. Seloignes, Ardennes, Belgium; Givetian. Specimen MRHN 1155, thin section a323. F, transverse section; G, longitudinal section. Scale bars represent 1 mm.

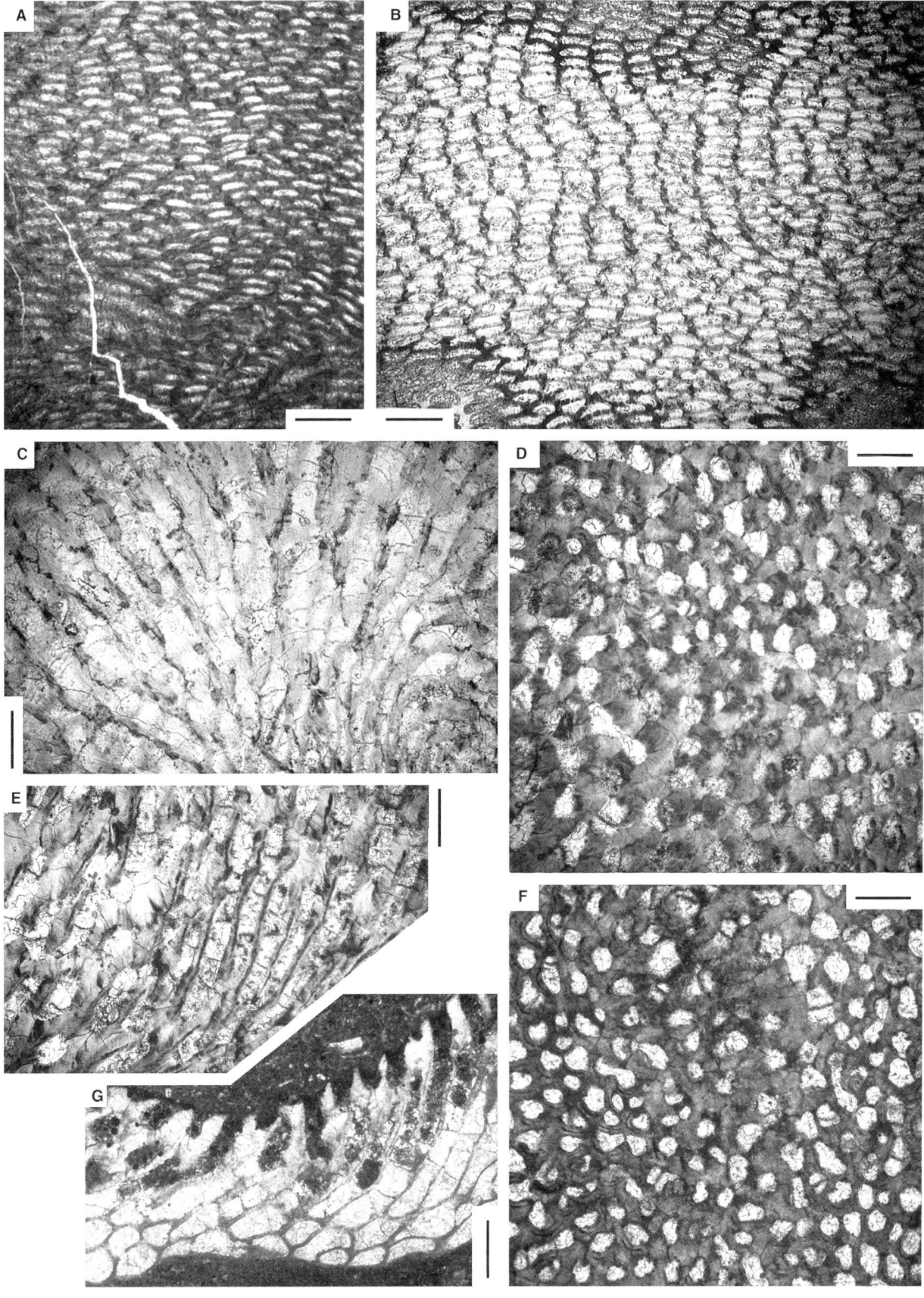

shaped, rarely irregular. Walls usually thin, but in several zones thicker, even. Median line visible on transverse sections as a discontinuous, dark, crescentic shape, with blurry edges. Connecting pores very rare (only one observed in the corallum). Tabulae thin, concave, sometimes feebly inclined or flat, spaced irregularly. Septal apparatus occur usually as a strong, massive, conical *hauptdorn*, rarely additional very small, sharp spines are visible.

Measurements. Table 12.

Remarks. The above-described corallum has been tentatively assigned to the genus *Alveolites* (on the basis of corallum shape, corallite shape and relation between corallite wall and corallite lumen), but the occurrence of rosebud-like arrangement of corallites and very rare connecting pores makes it similar to representatives of *Roseoporella*.

The discussed species resembles strongly *A. parvus*, mainly by corallite lumen diameters, but differs from this species by very rare connecting pores and unusually large spaces between tabulae. Shape of corallites makes it similar to *A. compressus*, from which it differs by much smaller corallites and large spaces between tabulae.

The intracolonial variation in *Alveolites* sp. is similar to that of the type species.

Occurrence. Kostomłoty Transitional Zone (Givetian).

Genus CRASSIALVEOLITES Sokolov, 1955

Type species. Alveolites crassiformis Sokolov, 1952 from the Givetian of the Kursk Region, Russia.

Remarks. The *Crassialveolites* is normally considered a valid genus (e.g. Dubatolov 1959; Sokolov 1962*a*; Kokscharskaya 1968; Nowiński 1976; Hill 1981), but Birenheide (1985) placed it into synonymy of *Alveolites*. Here, the genus is considered to be valid.

The species of *Crassialveolites* introduced by Deng (1979) and Tchi (1980, 1987) could not be compared with the species described here as the former are described in Chinese, which is not in agreement with the

recommendation 13B of the ICZN. Moreover, it appears that a diagnosis or description, derivation of the name, type locality and horizon are missing (or at least one of them); therefore, the validity of these specific names is questionable.

Crassialveolites cavernosus (Lecompte, 1933)
Figure 13D–G

*v.	1933	*Alveolites maillieuxi* var. *cavernosa* Lecompte, p. 38, pl. 3, fig. 4.
v.	1939	*Alveolites cavernosus* Lecompte; Lecompte, p. 45, pl. 7, figs 1–2.
?	1959	*Crassialveolites cavernosus* (Lecompte); Dubatolov p. 152, pl. 48, fig. 2.
?	1967	*Crassialveolites cavernosus* (Lecompte); Tong Dzuy, p. 112, pl. 22, fig. 4.
	1975	*Crassialveolites cavernosus*; Brice, Bigey, Mistiaen, Poncet and Rohart, p. 145.
p	1992	*Crassialveolites cavernosus* (Lecompte); Nowiński, p. 196, text-fig. 7C–D.
non	1996	*Alveolites cavernosus* Lecompte; Brühl, p. 18, pl. 7, fig. 22.
p	2003	*Crassialveolites cavernosus* (Lecompte); Nowiński, p. 148, pl. 94, figs 1–2.
	2007	*Crassialveolites cavernosus* Lecompte; Hubert, Zapalski, Nicollin, Mistiaen and Brice, p. 249.

Type material. Specimen MRHN 1155, thin section a323 (lectotype) and paralectotypes, from Seloignes 826b, Ardennes, Givetian (Lecompte 1933, pl. 3, fig. 4).

Material. Grabina (set C): one specimen, three thin sections (ZPAL T.25 GB 001–003; determination doubtful owing to obliquity of sections); Sowie Górki: two coralla, five thin sections (ZPAL T.25 SG 104–108); Posłowice: one corallum, two thin sections (ZPAL T.25 PW 029–030).

Description of the lectotype. Corallum massive, 120 mm of the largest diameter. Corallites twisted. Corallite lumina strongly rounded in cross section, round, oval, rarely subrectangular or irregular. In some places, numerous connecting pores make the corallum meandroidal. Walls thick, even, but the thickness vary

TABLE 12. Intracolonial variation in *Alveolites?* sp.

	Max LD	Min LD	DWT	PD	PS	TS
Specimen ZPAL T.25 LAS 100, from Laskowa Quarry, Givetian						
Mean	0.596	0.323	0.094	–	–	0.881
Standard deviation	0.0914	0.0542	0.0335	–	–	0.3413
N	31	31	31	–	–	31
Minimal value	0.46	0.20	0.04	–	–	0.48
Maximal value	0.84	0.42	0.18	–	–	2.00
V	0.153	0.168	0.356	–	–	0.387

Measurements in millimetres. For abbreviations, see Table 1.

TABLE 13. Intracolonial variation in *Crassialveolites cavernosus* (Lecompte).

	Max LD	Min LD	DWT	PD	PS	TS
Specimen MRHN 1155, thin section a323, from Seloignes 826b, Ardennes, Givetian; lectotype						
Mean	0.543	0.351	0.195	–	–	0.428
Standard deviation	0.1089	0.0672	0.0548	–	–	0.1369
N	30	30	30	–	–	30
Minimal value	0.28	0.26	0.08	–	–	0.22
Maximal value	0.72	0.50	0.28	0.20	–	0.76
V	0.200	0.191	0.280	–	–	0.320

Measurements in millimetres. For abbreviations, see Table 1.

TABLE 14. Intracolonial variation in *Crassialveolites cavernosus* (Lecompte).

	Max LD	Min LD	DWT	PD	PS	TS
Specimen ZPAL T.25 PW 029 30, from Posłowice, Frasnian						
Mean	0.572	0.407	0.231	–	–	–
Standard deviation	0.1289	0.0568	0.0649	–	–	–
N	23	23	23	–	–	–
Minimal value	0.30	0.24	0.12	–	–	–
Maximal value	0.80	0.52	0.36	0.18–0.20	–	–
V	0.225	0.139	0.281	–	–	–

Measurements in millimetres. For abbreviations, see Table 1.

strongly. Median line invisible. Tabulae probably numerous, flat or feebly concave, complete. Pores numerous, often closed by poral plate (none of the measurements was taken on the longitudinal section; thus, the value given in Table 13 is only an approximate). Septal spines rare, very small, conical, sharp.

Measurements. Table 13.

Description of the material from the Holy Cross Mts. Coralla probably bulbous, 60 mm of the largest diameter (size deduced from the thin section). Corallites twisted, irregularly bent. Corallite lumina rounded in cross section, oval, often strongly irregular, usually feebly elongated. The abundance of connecting pores causes often meandroidal character of the corallum. Walls thick, even, but the thickness vary strongly. Median line invisible. Tabulae numerous, flat or feebly concave, rarely inclined, most often complete. Pores numerous, usually closed by poral (measurements taken on cross section, therefore only the largest value is given in parentheses). Septal spines very rare, very small, blunt.

Measurements. Table 14.

Discussion. The material from the Holy Cross Mountains shows remarkable similarity to the lectotype with the exception that Polish specimens have more often irregular cross section of the corallite lumina. The meandroidal structure seems to be characteristic for this species. Slightly larger values of the minimal lumen width and double wall thickness are probably effect of incorrect cut-

ting. All the material is represented by sections of very small and fragmented specimens, usually cut at certain angle; hence, the study of intracolonial variation was undertaken only on one section.

The lectotype, similarly as in *A. parvus* does not permit making numerous necessary observations because pore diameter, pore spacing, tabulae morphology (however some features are visible on the investigated section) and finally tabulae spacing are not visible.

Occurrence. Northern Kielce Subregion (Frasnian), Southern Kielce Subregion (Late Givetian?–?Frasnian), Boulonnais (Givetian), Ardennes (Givetian), Kuznetsk Basin (Givetian).

Crassialveolites crassus (Lecompte, 1939)
Figures 14A–B, 15A–B

*v. 1939   *Alveolites crassus* nov. sp. Lecompte, p. 46, pl. 8 figs 1–2.

1952   *Alveolites crassus* Lecompte; Sokolov, p. 83, pl. 20, figs 1–3.

? v 1953   *Alveolites crassus* Lecompte; Stasińska, p. 222, text-fig. 7, pl. 1, fig. 3.

1959   *Crassialveolites crassus* (Lecompte); Dubatolov p. 148; pl. 49, figs 2, 3A–B, 4A–B.

? 1971   *Crassialveolites crassus multiaculatus* ssp. n. Dubatolov *in* Dubatolov and Spasski, p. 57, pl. 12, figs 1–5.

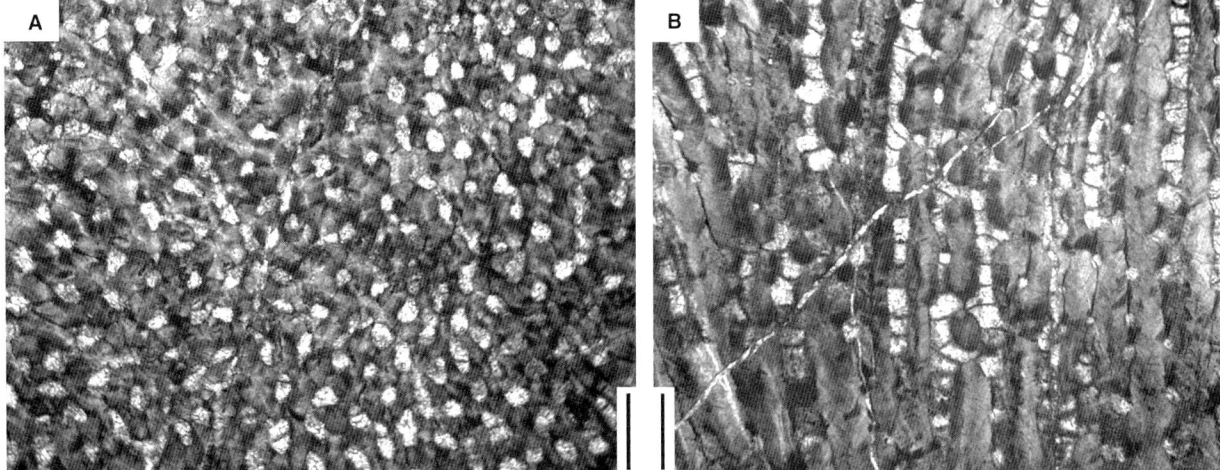

FIG. 14. A–B, *Crassialveolites crassus* (Lecompte, 1939), lectotype. Couvin, Belgium; Givetian, Gid (local stratigraphy). Specimen MRHN 90α (h), thin section a404. G, transverse section. H, longitudinal section. Scale bars represent 1 mm.

1972 *Crassialveolites crassus* (Lecompte); Yanet, p. 77, pl. 24, fig. 1.
? 1974 *Crassialveolites crassus* (Lecompte); Hladil, p. 219, pl. 1, fig. 1, pl. 2, fig. 2.
1975 *Crassialveolites crassus* (Lecompte); Khaiznikova, p. 72, pl. 14, figs 1–2.
? 1975 *Crassialveolites* cf. *crassus*; Brice, Bigey, Mistiaen, Poncet and Rohart, p. 145.
? 1984*b* *Crassialveolites crassus* (Lecompte); Hladil, p. 251, pl. 2, fig. 1.
non 1985 *Alveolites crassus* Lecompte; Birenheide, p. 81, pl. 25 fig. 1.
non 1988 *Crassialveolites crassus* (Lecompte); Tong-Dzuy, Nguyen and Kromykh, p. 89, pl. 38, figs 1, 3.
? 1992 *Crassialveolites multiperforatus* (Lecompte); Baikuchkarov, p. 107, text-fig. 1.
non 1993*a* *Alveolites* (*Crassialveolites*) *crassus* Lecompte; May, p. 158, pl. 7, fig. 2.
? 1993 *Crassialveolites crassus* (Lecompte); Lütte, p. 62, pl. 2, fig. 12.
? 1996 *Alveolites crassus* Lecompte; Brühl, p. 17, pl. 7, fig. 20, pl. 11, fig. 34.
vp 2003 *Crassialveolites crassus* (Lecompte); Nowiński, p. 149, pl. 94, figs 3–4.
vp 2003 *Crassialveolites multiperforatus* (Salée); Nowiński, p. 150, pl. 95, fig. 4, pl. 96, figs 4–5.
. 2007 *Crassialveolites crassus* Lecompte; Hubert, Zapalski, Nicollin, Mistiaen and Brice, p. 249.

Type material. Specimen MRHN 90α (h), thin section a404 (lectotype) and paralectotypes, from Couvin 6151d, Belgium; Givetian (Lecompte 1939, pl. 8, fig. 1).

Material. Czarnów: One corallum, two thin sections (ZPAL T.25 C010–011); Góra Cmentarna: one corallum, one thin section (ZPAL T.25 GC014); Posłowice: five coralla, 13 thin sections (ZPAL T.25: PW 001–002, 003a–b, 004–009, 017–020); Psie Górki: two coralla, two thin sections (ZPAL T.25: Sta PG 101, 102); Sowie Gorki: One corallum, four thin sections (ZPAL T.25 SWG 190–193); Wietrznia: three coralla, three thin sections (ZPAL T.25 Sta W 109–111).

Description of the lectotype. Corallum laminar aberrant, 17–28 mm in thickness, about 130 mm in maximal diameter (size deduced from the thin section). Corallites slightly curved, moderately long. Calyces irregularly deep, with sharp edges. Corallite lumina very variable in cross section, from nearly round, oval, subtriangular and subrectangular to irregular. Walls thick, with even surface but unevenly thickened, much thinner near connecting pores. Median line visible as thin, sharply delimitated, dark line, present only on some longitudinal sections. In the proximal parts of corallum corallite lumina of the first corallite layer are large and round (measuring up to 0.60 mm of the largest diameter), with thin walls. Tabulae thin, most often complete, rarely incomplete; flat, concave, rarely convex or inclined; numerous. Pores small, round, very often closed by poral plate. Septal apparatus very poorly developed, isolated short, conical spines occur irregularly throughout the corallum.

Measurements. Table 15.

Description of the material from the Holy Cross Mts. Coralla laminar aberrant or massive aberrant, about 60 mm in maximal diameter (size deduced from the thin section). Coral-

FIG. 15. A–B. *Crassialveolites crassus* (Lecompte, 1939). Czarnów, Holy Cross Mountains, Poland; set A (?), Givetian. Specimen ZPAL T.25 C010–011. A, transverse section; B, longitudinal section. C–D. *Crassialveolites* sp. Sowie Górki, Holy Cross Mountains, Poland; set B, Givetian, ZPAL T.25 SG 050–052. C, transverse section, D, longitudinal section. E–F. *Crassialveolites* aff. *crassus* (Lecompte, 1939). Posłowice; Holy Cross Mountains, Poland; Givetian. Specimen ZPAL T.25 PW 013–014. E, transverse section; F, longitudinal section. Scale bars represent 1 mm.

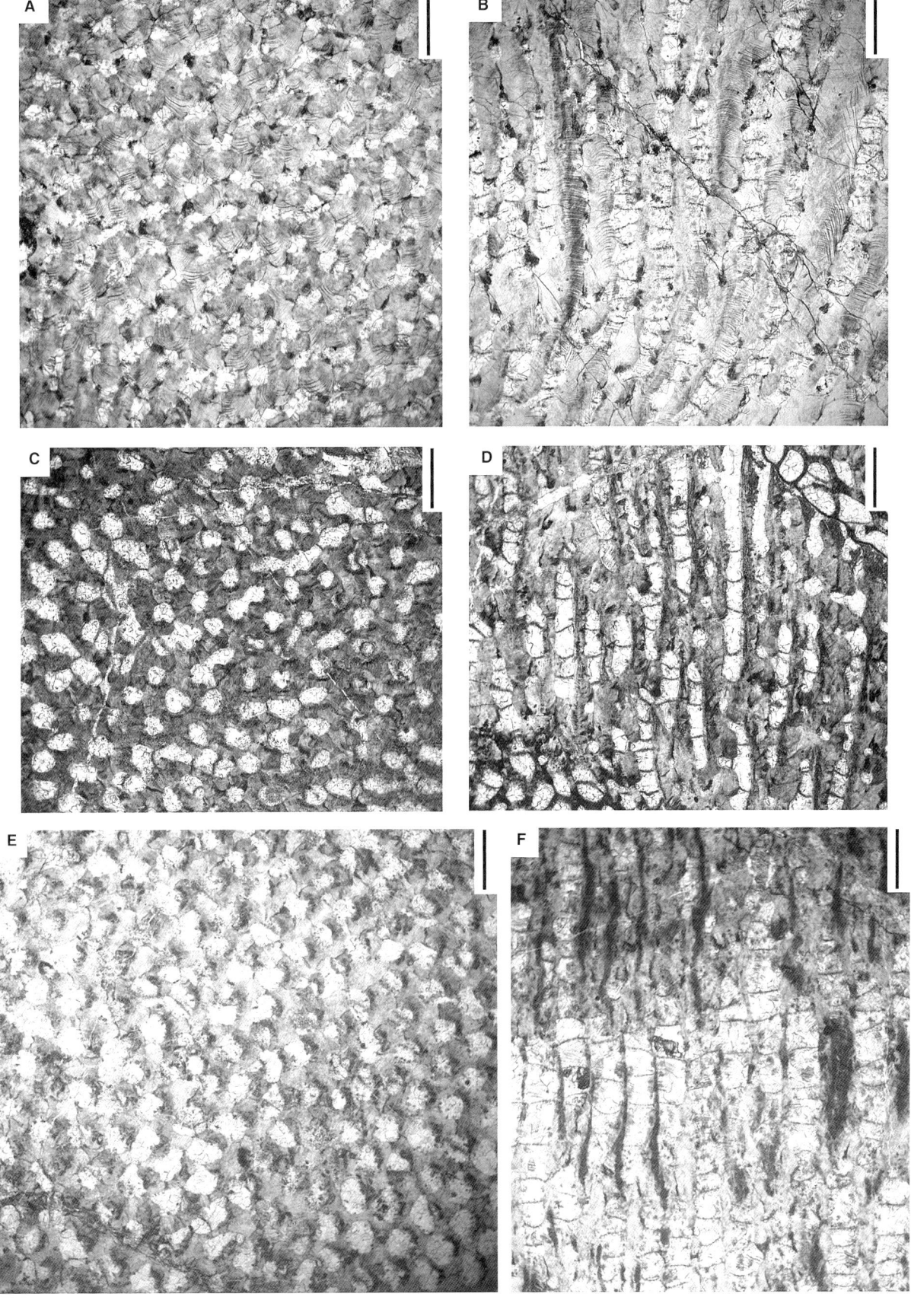

TABLE 15. Intracolonial variation in *Crassialveolites crassus* (Lecompte).

	Max LD	Min LD	DWT	PD	PS	TS
Specimen MRHN 90α (h), thin section a404, from Couvin 6151d, Givetian; lectotype						
Mean	0.400	0.256	0.338	0.132	0.554	0.333
Standard deviation	0.0854	0.0634	0.1004	0.0255	0.1734	0.0863
N	34	34	34	33	30	32
Minimal value	0.26	0.18	0.12	0.06	0.30	0.18
Maximal value	0.60	0.38	0.60	0.14	1.08	0.60
V	0.214	0.248	0.297	0.193	0.313	0.259

Measurements in millimetres. For abbreviations, see Table 1.

TABLE 16. Intracolonial variation in *Crassialveolites crassus* (Lecompte).

	Max LD	Min LD	DWT	PD	PS	TS
Specimen ZPAL T.25 C010–011, from Czarnów, Givetian						
Mean	0.479	0.293	0.311	0.112	–	0.373
Standard deviation	0.0769	0.0429	0.0638	0.0207	–	0.0841
N	28	28	28	19	7	30
Minimal value	0.32	0.20	0.14	0.08	0.40	0.26
Maximal value	0.70	0.40	0.46	0.14	0.66	0.54
V	0.160	0.147	0.205	0.185	–	0.225

Measurements in millimetres. For abbreviations, see Table 1.

lites slightly curved or straight. Corallite lumina very variable in cross section, from nearly round, oval, subtriangular and subrectangular to irregular and often elongated. The Frasnian material seems to have more frequent connecting pores. Such an abundance in some parts of coralla causes meandroid structure of the corallum. Walls thick, with uneven surface and unevenly thickened, much thinner near connecting pores. Median line most often invisible, blurry, discontinuous. In the proximal parts of corallum corallite lumina of the first corallite layer are large and round with thin walls. Tabulae thin, most often complete, rarely incomplete; flat, concave, rarely convex or inclined; they are numerous, distributed rather regularly. Pores small, round, sometimes closed by poral plate. Septal apparatus moderately developed, isolated short, conical spines occur irregularly throughout the corallum.

Measurements. Table 16.

Remarks. This species, comparing to other alveolitids, has different pattern of intracolonial variability: in the holotype, the corallite lumen diameters are more variable than in other alveolitids, while the tabulae spacing is less variable than in most of members of the family. The low variability of the diameters of the specimen from Czarnów can be explained by (1) the smaller section than in the case of Lecompte's type; in the lectotype, the measurements were taken from different parts of the section; (2) smaller number of measurements taken of

the Czarnów corallum; and (3) feebly smaller intracolonial variation.

For comparison with *C.? multiperforatus*, see the discussion on the latter species. It must be emphasized that the material from the Holy Cross Mts. is poorly preserved and only one corallum was measured in detail.

Occurrence. Northern and Central Kielce Subregions (Givetian–Frasnian); Eifel Mts. (Eifelian-Givetian), Ardennes (Givetian), Boulonnais (?; Frasnian), Moravia (Givetian), Kuznetsk basin (Givetian).

Crassialveolites? multiperforatus (Salée *in* Lecompte, 1933)
Figure 16A–D

* v. 1933 *Alveolites multiperforatus* sp. n. Lecompte, p. 39, pl. 3, fig. 1.
non 1952 *Alveolites multiperforatus* Salée; Sokolov, p. 82, pl. 19, figs 3–6.
vp 1953 *Alveolites multiperforatus* Lecompte; Stasińska, p. 228, text-fig. 11, ?pl. 2, fig. 3.
? 1959 *Alveolites multiperforatus* Salée; Dubatolov p. 140; pl. 47, fig. 1.
1972 *Crassialveolites multiperforatus* (Salée); Yanet, p. 75, text-fig. 11, pl. 23, figs 1–2.
? 1992 *Crassialveolites multiperforatus* (Lecompte); Baikuchkarov, p. 107, text-fig. 1.
vp 2003 *Crassialveolites multiperforatus* (Salée); Nowiński, p. 150, pl. 95, fig. 4, pl. 96, figs 4–5.

. 2007 *Crassialveolites multiperforatus* Salée; Hubert, Zapalski, Nicollin, Mistiaen and Brice, p. 250.

Type material. MRHN 4100a, thin section a324 (lectotype) and paralectotypes, from Chênée, Belgium; Frasnian (Lecompte 1933, pl. 3, fig. 1a).

Material. Góra Cmentarna: one corallum, three thin sections (ZPAL T.25 GC 004–006); Posłowice: two coralla corallum, nine thin sections (ZPAL T.25 PWa 017–020, PW 020–024); Sowie Górki: one corallum, two thin sections (ZPAL T.25 SG A01–A02); Stokówka (set C): two coralla, five thin sections (ZPAL T.25 ST 001–005); Wietrznia: seven coralla, nine thin sections (ZPAL T.25 Sta W 100–108) and the lectotype.

Description of the lectotype. Small, irregular corallum, measuring about 55 mm of the largest diameter (size deduced from thin section). Corallites slightly curved, in parts of corallum irregularly meandering and twisted, moderately long. Calyces deep, with sharp edges. Corallite lumina in cross section usually polygonal with rounded corners, oval, subtriangular or irregular. Walls moderately thick, even. Median line invisible. Tabulae very thin, most often complete, rarely incomplete; flat, concave, rarely inclined; numerous. Pores numerous, large, round, rarely closed by poral plate; they are placed in the corners but sometimes also on the walls. Septal apparatus developed in certain regions of corallum, mostly as small, button-like, blunt spines.

Measurements. Table 17.

Description of the material from the Holy Cross Mts. Small, lamellar coralla, measuring probably more than 70 mm of the largest diameter (size deduced from thin section). Corallites slightly curved, moderately long. Calyces deep, with sharp edges. Corallite lumina in cross section usually polygonal with strongly rounded corners, oval, subtriangular or irregular. Walls moderately thick, even. Median line invisible. Tabulae very thin, most often complete, rarely incomplete; flat, concave, rarely inclined; numerous. Pores numerous, large, round, sometimes closed by poral plate. Septal apparatus developed in certain regions of corallum, mostly as small, conical, sharp spines.

Measurements. Table 18.

Discussion. As it was already stated by Lecompte (1933, p. 11), this species shows some features making it close to favositids: polygonal sections of corallites (instead of typical for alveolitids elongated sections), and mural pores occurring sometimes also on walls (not in the corners; the latter feature makes this species close to *Scharkovaelites*). It is temporarily assigned here to *Crassialveolites*, basing on irregular thickening of the walls, and invisible median line (which is usually well visible in majority of favositids).

Baikuchkarov (1992) revised the collections of Sokolov, Dubatolov and Yanet and concluded that *C. crassus*, *C. crassiformis*, *C. domrachevi* are in fact conspecific with *C. multiperforatus*. The present author strongly disagrees with such an opinion, as, for example, the corallite diameter ranges of *C. crassus* and *C. crassiformis* are nonoverlapping (0.26–0.60 mm for *C. crassus* and 0.7–1.0 mm for *C. crassiformis*); similar situation is with *C. domrachevi* (1.0–1.2 mm; see Baikuchkarov 1992, table 1). Also, the study on the type specimens of both *C. crassus* and *C. multiperforatus* shows that these species are clearly distinct by several characters, such as lumen diameters, pore diameters and double wall thickness. Figure 17 shows the comparison of the double wall thickness in the lectotypes of two discussed taxa. The mean value of the double wall thickness is different for both species; moreover, the ranges of mean values ± standard deviation do not overlap. Similar, but not that clear situation is with maximal lumen diameter in lectotypes of *C. crassus* and *C. multiperforatus*, where their confidence ranges do not overlap (see Table 19).

Therefore, the two Lecompte's species are valid, while *C. crassiformis* Sokolov and *C. domrachevi* Dubatolov require separate study; it seems however that they are distinct taxa.

Occurrence. Northern, Central and Southern Kielce Subregions (Givetian–Frasnian); Ardennes (Frasnian); western slopes of Ural Mts. (Givetian), Kuznetsk Basin (Givetian).

Crassialveolites oliveri sp. nov.
Figure 18A–F

Derivation of the name. In honour of the late William A. Oliver, coral researcher.

Type strata. Biohermal limestones of Kowala Quarry (set ?C), below the *punctata* Zone, Early Frasnian.

Type locality. Kowala Quarry, Holy Cross Mountains, Poland.

Holotype. ZPAL T.25 MZHT210/D (Fig. 18A–F); one very well-preserved corallum; four thin sections, one ultrathin section.

Type horizon. Kowala Quarry, western wall of the quarry, biohermal equivalent of the biostromal set A of the railroad cut.

Diagnosis. Coralla laminar. Corallites very variable in shape, nearly round, oval, lenticular, subtriangular, irregularly polygonal. The size of their lumen variable, 0.38 (±0.05) × 0.53 (±0.09) mm. Walls vertically and horizontally uneven, double wall thickness 0.29 (±0.06) mm. Connecting pores 0.08–0.16 mm in diameter, often closed by poral plate. Tabulae concave and irregularly folded, spaced 0.39 (±0.14) mm. Septal spines numerous.

Description. Laminar, encrusting corallum, measuring 7–10 × 80 × 105 mm. Calyces shallow, polygonal with rounded

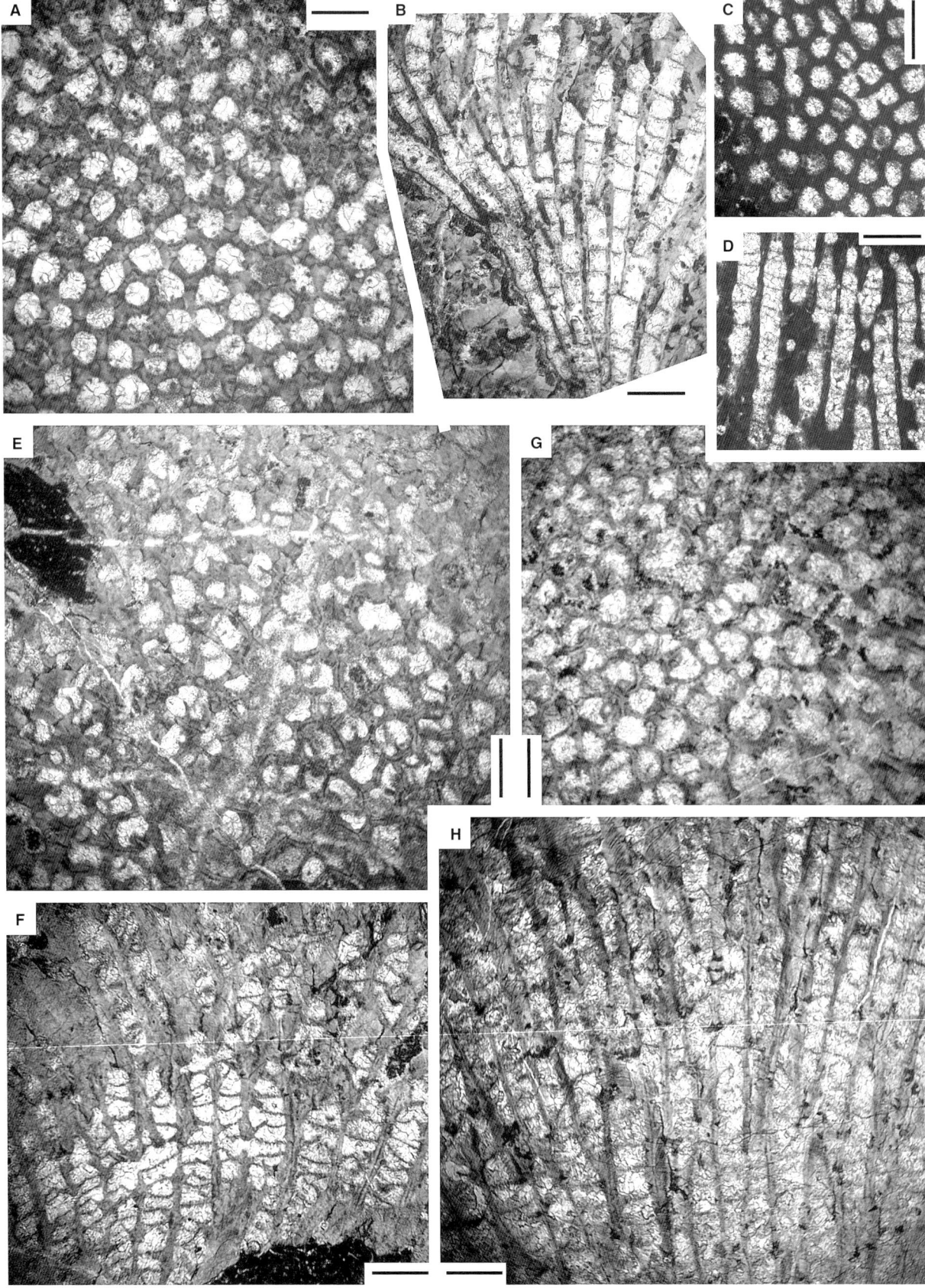

TABLE 17. Intracolonial variation in *Crassialveolites*? *multiperforatus* (Lecompte).

	Max LD	Min LD	DWT	PD	PS	TS
Specimen MRHN 4100a, thin section a324, from Chênée, Frasnian; lectotype						
Mean	0.518	0.427	0.152	0.185	0.591	0.341
Standard deviation	0.0792	0.0634	0.0482	0.0269	0.1177	0.0791
N	33	33	33	33	32	33
Minimal value	0.30	0.40	0.08	0.12	0.40	0.22
Maximal value	0.58	0.76	0.28	0.24	0.94	0.56
V	0.153	0.149	0.318	0.146	0.199	0.232

Measurements in millimetres. For abbreviations, see Table 1.

TABLE 18. Intracolonial variation in *Crassialveolites*? *multiperforatus* (Lecompte).

	Max LD	Min LD	DWT	PD	PS	TS
Specimen ZPAL T.25 ST 001–003 from Stokówka, late Givetian						
Mean	0.545	0.366	0.216	0.174	–	0.468
Standard deviation	0.1025	0.0597	0.0621	0.0250	–	0.1806
N	32	32	32	23	8	32
Minimal value	0.32	0.22	0.10	0.14	0.56	0.10
Maximal value	0.74	0.5	0.34	0.22	0.86	0.84
V	0.188	0.163	0.288	0.143	–	0.386
Specimen ZPAL T.25 PW 020–024, from Posłowice, Givetian						
Mean	0.616	0.422	0.189	0.170	–	0.529
Standard deviation	0.0781	0.0572	0.0577	0.0276	–	0.1470
N	34	34	34	12	6	33
Minimal value	0.46	0.34	0.10	0.12	0.68	0.30
Maximal value	0.78	0.60	0.30	0.20	0.74	0.80
V	0.127	0.135	0.305	0.163	–	0.278

Measurements in millimetres. For abbreviations, see Table 1.

corners. Corallites in the transverse section very variable, from nearly round or oval, lenticular, subtriangular, to irregularly polygonal, rarely slightly crescentic. Corallite walls thick, uneven, thinned near connecting pores. Median line visible only on longitudinal sections, as dark, uneven, discontinuous line. Some of them are menisc-shaped, thickened in the middle; the others are flat and even. In few places, numerous pores cause change in cerioidal character of the colony into slightly meandroidal; such a feature occurs only locally in certain parts of the corallum. Tabulae irregular, often concave or folded, slightly inclined, spaced irregularly. Connecting pores round and irregular, numerous, distributed irregularly. The pores are often closed by convex or straight poral plates, 0.02–0.04 mm in thickness (Fig. 18C–D). Septal spines from one to eight in one cross section of the corallite, rarely absent. They are slim, thin at the base, long, usually sharp-pointed; the ratio width at the base/length is usually like 2:1, most often 0.08–0.10 mm in length.

Measurements. Table 20.

Discussion. The description of a new species on a single specimen is always risky. In the case of *C. oliveri* sp. n. always at least one feature is entirely different from the other known species of this genus; thus, it has been decided to create a new taxon.

Comparisons with other species of *Crassialveolites* are given in Table 21.

The new species differs from *C. crassus* (Lecompte) by having larger corallite diameters and slightly thinner walls

FIG. 16. A–B, *Crassialveolites*? *multiperforatus* (Lecompte, 1939). Posłowice; Holy Cross Mountains, Poland; Givetian. ZPAL T.25. PW 020–024. A, transverse section; B, longitudinal section; C–D, *Crassialveolites*? *multiperforatus* (Lecompte, 1939), lectotype. Chênée, Belgium; Frasnian, F2 II (local stratigraphy). Specimen MRHN 4100a, thin section a324. C, transverse section; D, longitudinal section. E–F. *Caliapora* (*Caliapora*) *battersbyi battersbyi* (Milne-Edwards and Haime, 1851). Czarnów, Holy Cross Mountains, Poland; Givetian. Specimen ZPAL T.25 C034–035. E, transverse section; F, longitudinal section. G–H, *Caliapora* (*Caliapora*) *venusta* Yanet, 1972. Czarnów, Holy Cross Mountains, Poland; Givetian. Specimen ZPAL T.25 C045/047. G, transverse section; H, longitudinal section. Scale bars represent 1 mm.

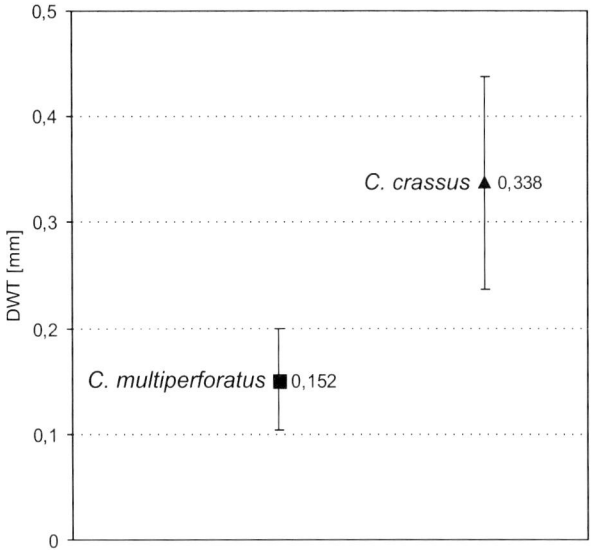

FIG. 17. Comparison of mean values (square and triangle) and standard deviation (whiskers) of double wall thickness in the lectotypes of *C. crassus* and *C. multiperforatus* ($N = 34$ and $N = 33$, respectively).

TABLE 19. Comparison of maximal lumen diameters in lectotypes of *C. crassus* and *C.? multiperforatus*.

	Mean	Standard deviation	95 per cent confidence
C. crassus, lectotype, $N = 34$	0.400	0.0854	(0.3702; 0.4298)
C. multiperforatus, lectotype, $N = 33$	0.518	0.0792	(0.4895; 0.5456)

Measurements in millimetres.

(Table 21), from *C. mirus* Dubatolov by having smaller pores, from *C. cavernosus* (Lecompte) both by smaller pores and by different shape of corallum (massive in *C. cavernosus*) and from *C. evidens* Dubatolov by having smaller corallite diameters and corallum shape (bulbous in *C. evidens*). From *C.? multiperforatus* (Lecompte), it differs in smaller number of smaller connecting pores, shape of corallum (massive in *C. multiperforatus*) and absence of mural pores. *C. polonicus* Nowiński has larger corallites and bulbo-cylindrical coralla. From *C. incrassatus*, Dubatolov differs in having much thinner walls and shape of corallum (bulbous and discoidal in *C. incrassatus*). *C. macrotrematus* Dubatolov has much larger connecting pores and thicker walls, while *C. krekovensis* Dubatolov has smaller corallite diameters and a different shape of corallites in cross section. From *C.* sp. described below, it differs by much more frequent elongated, irregular (in cross section) corallites, uneven walls and more frequent pores (however, all the biometrical values remain

similar). Table 21 does not include *C. tomskoensis* Dubatolov (Dubatolov *et al.* 1968, p. 101, pl. 47, figs 1–2), neither *C.(?) diversus* Mironova (Mironova, 1974, p. 86, pl. 64, fig. 1) because and in the opinion of this author these species should not be referred to the genus *Crassialveolites*.

Occurrence. Southern Kielce Subregion (Early Frasnian).

Crassialveolites aff. *crassus* Lecompte, 1939
Figure 15E–F

? v 1953	*Alveolites crassus* Lecompte; Stasińska, p. 222, text-fig. 7, pl. 1, fig. 3.
? 1967	*Crassialveolites crassus* (Lecompte); Tong Dzuy, p. 114, pl. 23, fig. 2.
? 1969b	*Crassialveolites crassus* (Lecompte); Stasińska, p. 772.
? 1986	*Crassialveolites crassus* (Lecompte); Nowiński and Prejbisz, p. 242.
? 1988	*Crassialveolites crassus* (Lecompte); Tong-Dzuy, Nguyen and Kromykh, p. 89, pl. 38, figs 1, 3.
? 1996	*Alveolites crassus* Lecompte; Brühl, p. 17, pl. 7, fig. 20, pl. 11, fig. 34.
vp 2003	*Alveolites obtortus* Lecompte; Nowiński, p. 143, non pl. 86, figs 1–2.
vp 2003	*Crassialveolites crassus* (Lecompte); Nowiński, p. 149, non pl. 94, figs 3–4.

Material. Laskowa (set ?A): one corallum, three thin sections (ZPAL T.25 L023–025); Posłowice: one corallum, seven thin sections (ZPAL T.25 PW 010–016).

Description. Corallum laminar aberrant, about 90 mm in maximal diameter (size deduced from the thin section). Corallites twisted, curved or straight. Corallite lumina very variable in cross section, from nearly round, oval, subtriangular and subrectangular to irregular and often elongated. Walls thick, with uneven surface and unevenly thickened, much thinner near connecting pores. Median line visible only on longitudinal sections, blurry, discontinuous. Tabulae thin, complete and often incomplete; flat, concave, sometimes strongly convex, rarely inclined; they are numerous, distributed rather regularly. Pores small, round, often closed by poral plate. Septal apparatus well developed, short, conical spines occur irregularly throughout the corallum.

Measurements. Table 22.

Remarks. The described above specimen shows some similarities to *C. crassus*, from which it differs by having (1) having larger diameters of corallite lumina; (2) thinner walls; (3) less densely spaced tabulae; and (4) better development of septal apparatus. The larger size of connecting pores may be caused by intersection of the thin

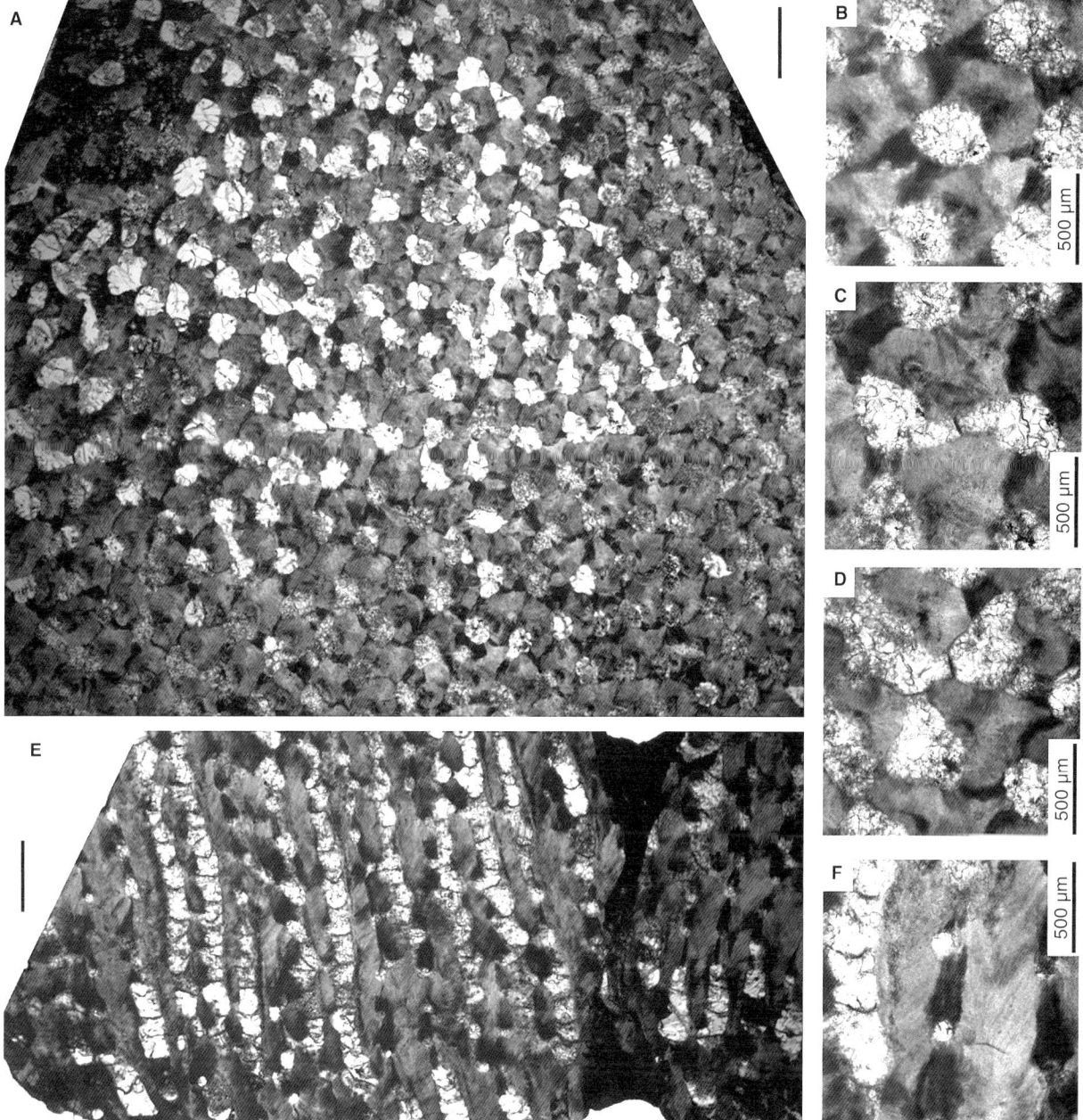

FIG. 18. A–F, *Crassialveolites oliveri* sp. nov., holotype. Kowala Quarry, Holy Cross Mountains, Poland; rubble of sets A–C, Early Frasnian. Specimen MZHT 210/D. A–D, transverse sections, photographs B–E show details of corallites. Note the poral plate on photograph C. E–F, transverse sections. See also Figure 23C. Scale bars represent 1 mm, if not stated otherwise.

section and wall. The material described by Tong-Dzuy (1967) from the Devonian of Vietnam and referred to *C. crassus* is very similar to *C.* aff. *crassus* described here, but the illustrations of the Viet Nam material do not show the diagnostic characters.

Occurrence. Kostomłoty Transitional Zone and Central Kielce Subregion (Givetian), Eifel Mts. (Eifelian), Pomerania (Givetian), Vietnam (Middle Devonian).

Crassialveolites sp.
Figure 15C–D

1992. *Crassialveolites evidens* Dubatolov; Nowiński, p. 196, fig. 8.
2003. *Crassialveolites evidens* Dubatolov; Nowiński, p. 149, pl. 95, figs 1–3.

TABLE 20. Intracolonial variation in *Crassialveolites oliveri* sp. nov.

	Max LD	Min LD	DWT	PD	PS	TS
Specimen ZPAL T.25 MZHT210/D, Kowala Quarry, Early Frasnian						
Mean	0.538	0.383	0.291	0.118	0.434	0.392
Standard deviation	0.0921	0.0513	0.0620	0.0229	0.0617	0.1360
N	44	44	38	32	24	32
Minimal value	0.24	0.40	0.16	0.06	0.30	0.20
Maximal value	0.50	0.76	0.46	0.16	0.60	0.80
V	0.171	0.134	0.213	0.194	0.142	0.347

Measurements in millimetres. For abbreviations, see Table 1.

Material. Czarnów (set B): One corallum, five thin sections (ZPAL T.25 C058–062): Sowie Górki (set B): Four coralla, 11 thin sections (ZPAL T.25 SG050–060).

Description. Coralla small, massive. Corallites long, usually straight, often weakly meandering. Corallite lumina usually round or nearly round, sometimes feebly oval, measuring 0.28–0.40 mm in diameter (if round) or 0.32–0.40 × 0.40–0.70 mm (if oval). Walls very thick even, 0.22–0.42 mm in thickness; median line is invisible. Tabulae numerous, concave, feebly inclined or folded, rarely with umbilical structure, spaced every 0.14–0.76 mm. Connecting pores numerous, round, sometimes faintly irregular, often closed by poral plate, about 0.12 mm in diameter. Septal spines rare, very small, conical, sharp, irregularly distributed. On the early ontogenetic stages, the corallites are polygonal, have large diameters and thin walls.

Discussion. The specimens described above were determined by Nowiński (1992) as *C. evidens* Dubatolov. However, they differ significantly by having larger lumina diameters and thicker walls (see Table 21 for data on *C. evidens*). The specimens are closest to the *C. inflatus* Khaiznikova, from which they differ by even walls (very uneven in the latter species, see Khaiznikova 1975, pl. 16, fig. 2) and predominance of oval lumina (mostly elongated in *C. inflatus*). It also resembles *C. ovlatschanus* Koksharskaya, from which it differs by smaller pores (0.17–0.20 mm in *C. ovlatschanus*; see Table 21).

Occurrence. Northern and Central Kielce Subregion (Middle–Late Givetian).

Genus ALVEOLITELLA Sokolov, 1952

Type species. Alveolitella fecunda (Salée *in* Lecompte, 1933) from the Givetian of the Ardennes.

Discussion. Alveolitella Sokolov (1952, p. 77) has been recognized as a junior subjective synonym of *Alveolites* by Hill (1981) and Birenheide (1985). However, most coral

TABLE 21. Comparison of species of *Crassialveolites*.

	Corallum shape	Corallite diameter	Wall thickness	Pore diameter	Data from
C. oliveri	L	0.24–0.50 × 0.40–0.76	0.16–0.46 (0.29)	0.06–0.16	This paper
C. abramovi	L, V	0.25–040	0.18–0.22	0.10–0.12	Dubatolov (1969)
C. cavernosus	M	0.26–0.50 × 0.28–0.72	0.08–0.28	(0.20)	This paper
C. crassus	L	0.18–0.38 × 0.26–0.60	0.12–0.60	0.06–0.14	This paper
C. crustaceus	M	0.50–0.70 × 0.60–0.90	0.25–0.30	0.10–0.15	Kokscharskaya (1968)
C. domrachevi	?	1.00	0.30–0.35	0.15–0.20	Dubatolov (1959)
C. evidens	B	0.25, 0.20 × 0.40	0.18–0.22	0.10–0.12	Dubatolov (1963)
C. incrassatus	B	0.25–0.55	0.18–0.30	0.15	Dubatolov (1963)
C. inflatus	M	0.3–0.5 × 0.5–0.7	0.30–0.50	0.10–0.15	Khaiznikova (1975)
C. krekovensis	I	0.30 × 0.50	0.10–0.25	0.10–0.12	Dubatolov (1959)
C. macrotrematus	M	0.25–0.40; 0.50–0.70	0.2–0.3	0.25–0.30	Dubatolov (1963)
C. mirus	M	0.30–0.40 × 0.40–0.60	0.15–0.25	0.2	Dubatolov (1959)
C. multiperforatus	M	0.30–0.58 × 0.40–0.76	0.08–0.28	0.12–0.24	This paper
C. ovlatschanus	M	0.30–0.48 × 0.50–0.70	0.10–0.30	0.17–0.20	Koksharskaya (1968)
C. polonicus	B, C	0,8–1,0	0.08–0.25	0.25–0.30	Nowiński (1976)
C. spiralis	M	0.27–0.30 × 0.30–0.60	(up to) 0.27	0.07–0.10	Koksharskaya (1968)
C. symbioticus	L	0.4–0.5 × 0.7	0.10–0.12	0.10–0.12	Dubatolov (1959)
C. sp.	M	0.32–0.40 × 0.40–0.70	0.22–0.42	0.10	This paper

B, bulbous; C, columnar; I, irregular; L, laminar; M, massive; V, variable. All values in millimetres.

TABLE 22. Intracolonial variation in *Crassialveolites* aff. *crassus* (Lecompte).

	Max LD	Min LD	DWT	PD	PS	TS
Specimen ZPAL T.25 PW 010–016, from Posłowice, Frasnian						
Mean	0.506	0.347	0.244	–	–	0.470
Standard deviation	0.0817	0.0569	0.0742	–	–	0.1877
N	31	31	31	6	–	31
Minimal value	0.28	0.28	0.10	0.10	–	0.14
Maximal value	0.66	0.50	0.36	0.20	–	1.04
V	0.161	0.164	0.304	–	–	0.399

Measurements in millimetres. For abbreviations, see Table 1.

researchers recognize the taxon as a distinct genus or as a subgenus of *Alveolites* (i.e. Nowiński 1976; Dubatolov and Ivanovski 1977; Iven 1980; Lin *et al.* 1988; May 1993*a*; Lukin 1998).

The characteristic feature of this genus is the branching corallum with polygonal corallites in the axial zone and alveolitoid (elongated) in peripheral zone. Moreover, species of *Alveolitella* do not possess a septal apparatus. For these reasons, *Alveolitella* is regarded here as a different and distinct genus.

Alveolitella fecunda (Salée *in* Lecompte, 1939)
Figure 19F–L

> *v. 1939 *Alveolites fecundus* (Salée); Lecompte, p. 57, pl. 9, figs 2–3.
>
> 1953 *Alveolites fecundus* (Salée); Stasińska, p. 225, pl. 1, fig. 4, pl. 2, fig. 1.
>
> 1959 *Alveolitella fecunda* (Salée); Dubatolov, p. 1960, pl. 52, fig. 4.
>
> ? 1969*b* *Alveolitella fecunda* (Salée); Stasińska, p. 772.
>
> 1972 *Alveolitella fecunda* (Salée); Yanet, p. 79, text-fig. 13, pl. 25, fig. 2.
>
> 1976 *Alveolitella fecunda* (Salée); Nowiński, p. 61, pl. 7, fig. 1.
>
> 1985 *Alveolites fecundus* (Lecompte); Birenheide, p. 80.
>
> 1986 *Alveolitella fecunda* (Salée); Nowiński and Prejbisz, p. 242.
>
> ? 1993 *Alveolitella fecunda* (Lecompte); Lütte, p. 62, pl. 2, figs 10–11, 13.
>
> v. 2003 *Alveolitella fecunda* (Lecompte); Nowiński, p. 146, pl. 90, figs 2–4.
>
> ? 2005 *Alveolites (Alveolitella) fecundus* Lecompte; Stadelmaier, Nose, May, Salerno, Schröder and Leinfelder, p. 7, pl. 1, figs 1–7.
>
> . 2007 *Alveolitella fecundus* Lecompte; Hubert, Zapalski, Nicollin, Mistiaen and Brice, p. 249.

Type material. Specimen MRHN a417 (coll. M. Lecompte), Bruxelles, Belgium (Lecompte 1939, pl. 9, figs 2–3).

Material. Laskowa Góra Quarry (set B): one corallum, two thin sections (ZPAL T.25 L210–211); Marzysz: six coralla, 14 thin sections (ZPAL T.25: M001–012, M015–016); Miedzianka: two coralla, seven thin sections (ZPAL T.25: MD 013–015, MD009–012); Posłowice: 15 coralla, 36 thin sections (ZPAL T.25: PL 001–011, P100–118, P120–123, 200–201, 210, 219); Trzemoszna: 16 coralla, 35 thin sections (ZPAL T.25: TR 001–TR021, TR028–041) and the lectotype.

Description of the lectotype. Corallum branching, the branch diameter varies between 17 and 20 mm. Corallites in the axial zone of branches polygonal, become flattened and feebly twisted towards the peripheral zone. In the passage between these zones, corallites become bent at about 60 degrees to the branch axis. The axial zone occupies 1/3–1/2 of the branch diameter. In the axial zone walls are thin and even, and they thicken significantly in the peripheral zone. Thin median line is often visible. Pores very abundant, round, spaced regularly. Tabulae numerous, concave and inclined, are spaced more densely in the peripheral zone. Septal apparatus absent.

Measurements. Table 23.

Description of the Holy Cross Mts. material. Coralla branching. Branches are very variable in size, ranging from 8 to 22 mm in diameter. Corallites in the axial zone of branches polygonal, they become flattened towards the peripheral zone. In the passage between these zones, corallites become bent at about 50–70 degrees to the branch axis. The axial zone occupies one-fourth to one-fifth of the branch diameter. It must be noted also that the diameter of axial zone varies only a little, between approximatively 3 and 5 mm; it seems to be constant throughout the branch (therefore, young branches have relatively large axial zones). In the axial zone, walls are thin and even, and they thicken significantly in the peripheral zone. Thin median line usually visible. Pores not numerous, round or oval, their spacing unknown. Tabulae numerous, concave, flat and inclined, spaced more densely in the peripheral zone. Septal apparatus absent.

Measurements. Table 24.

Regeneration phenomena and teratology. In the corallum from Marzysz (ZPAL T.25 M001–003), possibly regeneration phenomena are visible. After a death of three individuals in the central zone of corallum, the neighbouring individuals overgrow the (probably) empty corallites (Fig. 19K). The cause of

TABLE 23. Intracolonial variation in *Alveolitella fecunda* (Lecompte).

	Max LD (axial zone)	DWT	PD	PS	TS (ax z)	TS (p z)
Specimen a417, from Durbuy, Gib, Givetian; Belgium. Lectotype						
Mean	0.435	0.141	0.147	0.650	0.965	0.399
Standard deviation	0.0821	0.0465	0.0264	0.2243	0.3212	0.1433
N	30	32	34	23	15	33
Minimal value	0.24	0.06	0.10	0.34	0.42	0.16
Maximal value	0.58	0.26	0.22	1.04	1.52	0.68
V	0.189	0.330	0.180	0.345	0.333	0.359

Measurements in millimetres. ax z, axial zone of branch; p z, peripheral zone of branch; for other abbreviations see Table 1.

death is unclear; nonetheless in the middle one of three over-grown individuals, last tabula is strongly convex – such tabulae do not occur in other places of this corallum, neither in other coralla. Newly appeared corallites have in the proximal parts much thinner walls than neighbouring ones. The sediment influx as a cause of death can be excluded, as these corallites are filled by sparite, feebly darker than in the neighbouring ones. One of multiple possibilities is mechanical damage, as in two places walls are broken (which may be an effect of recrys-tallization as well).

Discussion. The specimens from the Holy Cross Moun-tains seem to be conspecific with the type material described by Lecompte (1933); however, they differ from the lectotype by having narrower axial zone of branches, somewhat thicker corallite walls and less abundant connecting pores. As mentioned in the description, the diameter of the axial zone is constant; consequently, the ratio axial/peripheral zone changes during the astogeny.

The intracolonial variation in this species remains on high and very high level, comparing to other alveolit-ids. Also intraspecific variation is relatively high, except for the corallite lumen diameter, which remains similar in all specimens, although within coralla this character is rather highly variable (variation coefficient is close to 0.2).

Two coralla coming from Trzemoszna (ZPAL T.25: TR034, 035; Fig. 30A–B) are infested by *Helicosalpinx* par-asites (see below).

Occurrence. Central Kielce Subregion (Middle–Late Givetian to Early Frasnian); Ardennes, Eifel Mts. (Givetian), western slopes of Ural Mts. (Givetian).

Alveolitella polygona Nowiński, 1992
Figures 19A–E

*v. 1992 *Alveolitella polygona* sp. n. Nowiński, p. 197, 205, text-fig. 9A–E.
v. 2003 *Alveolitella polygona* Nowiński; Nowiński, p. 147, pl. 91, figs 4–7.

Type material. Specimen ZPAL T XVIII-18/51 (thin sections PW 146–149), Institute of Paleobiology PAS, Warsaw (Nowiński 1992, text-fig. 2A–B).

Material. Posłowice (set B): two coralla, 10 thin sections (ZPAL T.25 PW 140–149).

Description. Coralla small, branching, round or somewhat oval in cross section, 4–7 mm of the largest diameter. Corallites in the axial zone polygonal, with large visceral chamber, rounded in the corners. Their lumen measures 0.24–0.44 mm of the larg-est diameter. Corallites become twisted towards the peripheral zone. The axial zone wide, occupying 70–80 per cent of the branch diameter. Corallite walls relatively thin in the axial zone (0.04–0.10 mm), and they slightly thicken towards the periphery (0.12–0.20 mm). Median line granular, discontinuous, well visi-ble in entire coralla. Tabulae flat, concave, sometimes folded, varying in thickness, but generally thin; spaced irregularly, every 0.14–0.60 mm. Connecting pores circular, rare, 0.10–0.14 mm in diameter, spaced irregularly.

Discussion. In addition to the remarks of Nowiński (1992, p. 206), it can be mentioned that the species differs from most of the other species of *Alveolitella* by having rela-tively thin branches of corallum. Only *Alveolitella karmak-ensis* (Tchernychev) from the Eifelian of Kuzbas has

FIG. 19. A–E, *Alveolitella polygona* Nowiński, 1992. Posłowice, Holy Cross Mountains, Poland; set B, Givetian. Specimen ZPAL T.25 PW 140–145. A–B, transverse section; C, longitudinal section; thin sections ZPAL T.25 PW 146–149 (holotype). D, transverse section. E, longitudinal section. F–G, *Alveolitella fecunda* (Salée *in* Lecompte, 1933), lectotype. Durbuy, Ardennes, Belgium. Givetian, 'Gib' of local stratigraphy. Thin section MRHN a417: F, transverse section; G, longitudinal section. H–I, *Alveolitella fecunda* (Salée *in* Lecompte, 1933). Laskowa Góra Quarry, Holy Cross Mountains, Poland; set B, Szydłówek Beds, Givetian. Specimen ZPAL T.25 L210–211. H, transverse section. I, longitudinal section. J–L, *Alveolitella fecunda* (Salée *in* Lecompte, 1933). Marzysz, Holy Cross Mountains, Poland; Givetian. Specimen ZPAL T.25 M 001–002. J, transverse section. K, longitudinal section. L, longitudinal section, detail of K showing regeneration phenomena. Scale bars represent 1 mm.

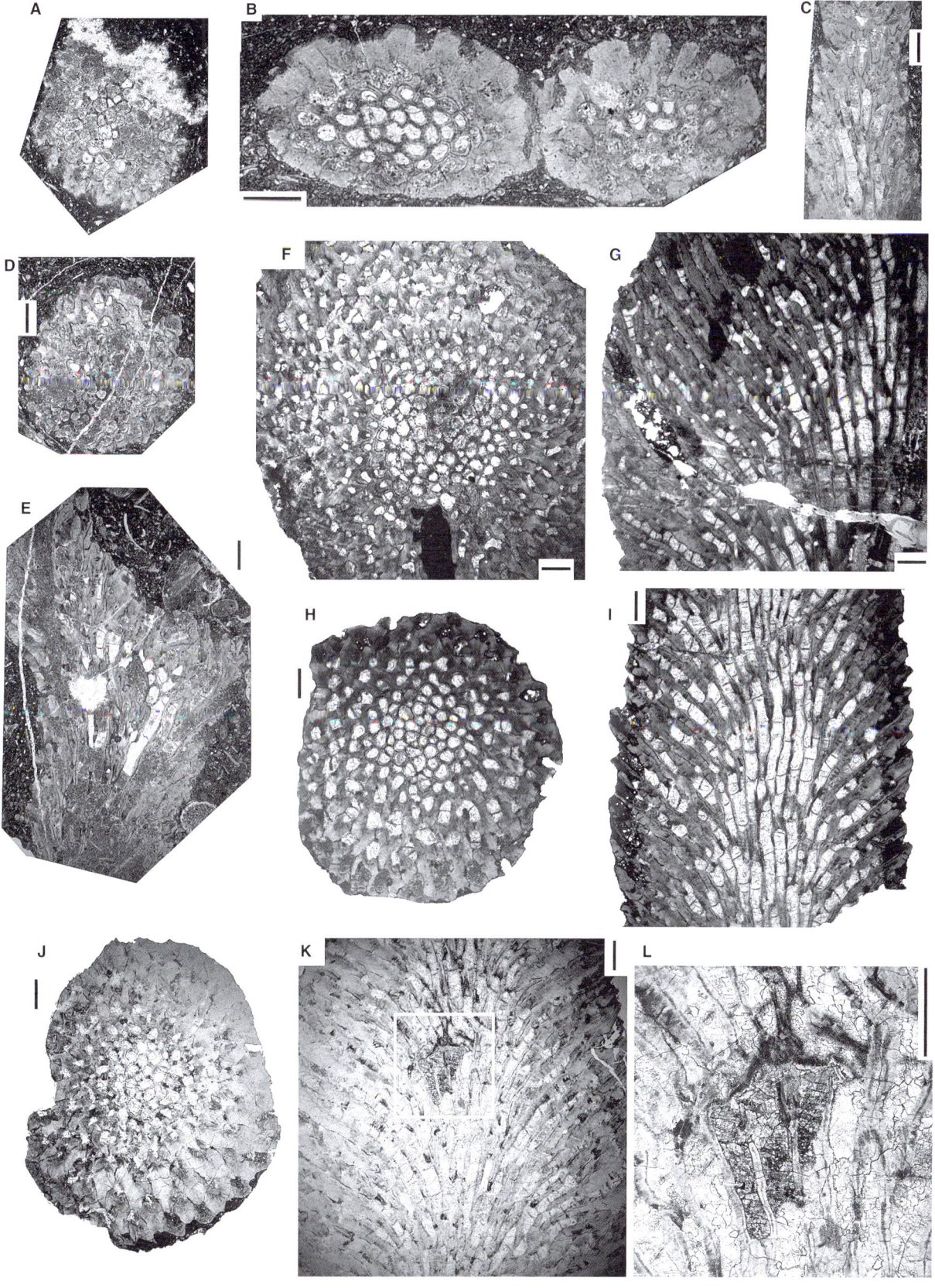

TABLE 24. Intracolonial variation in *Alveolitella fecunda* (Lecompte).

	Max LD (axial zone)	DWT	PD	PS	TS (ax z)	TS (p z)
Specimen ZPAL T.25 TR002–003, from Trzemoszna, Givetian						
Mean	0.426	0.137	–	–	–	0.445
Standard deviation	0.0736	0.0215	–	–	–	0.1294
N	32	32	7	–	2	32
Minimal value	0.26	0.08	0.10	–	0.40	0.24
Maximal value	0.54	0.18	0.16	–	1.00	0.74
V	0.173	0.156	–	–	–	0.291
Specimen ZPAL T.25 PL009–011 from Posłowice, Givetian						
Mean	0.441	0.233	0.164	–	0.924	0.478
Standard deviation	0.0912	0.0576	0.0266	–	0.4521	0.1778
N	28	30	16	–	21	30
Minimal value	0.24	0.06	0.12	–	0.20	0.14
Maximal value	0.58	0.26	0.20	–	2.00	0.80
V	0.207	0.247	0.162	–	0.489	0.372
Specimen ZPAL T.25 M001–003 from Marzysz, Givetian						
Mean	0.387	0.132	–	–	–	0.500
Standard deviation	0.0742	0.0238	–	–	–	0.1235
N	30	30	6	3	3	14
Minimal value	0.24	0.08	0.10	0.62	0.4	0.32
Maximal value	0.56	0.16	0.14	0.78	0.66	0.70
V	0.192	0.181	–	–	–	0.247

Measurements in millimetres. ax z, axial zone of branch; p z, peripheral zone of branch; for other abbreviations see Table 1.

similar branch diameters, but it differs in the wall thickness, which is thicker in *A. karmakensis*, and much larger corallite diameters.

Occurrence. Central Kielce Subregion (Givetian).

Family CALIAPORIDAE Mironova, 1974

Type genus. Caliapora Schlüter, 1889.

Remarks. Hill (1981) placed the subfamily Caliaporinae within the Alveolitidae, but Fernández-Martínez and Tourneur (1993) followed by Iven *et al.* (1997) moved its rank to family level. The present author follows the systematics of these authors and regards Caliaporidae as separate family.

Genus CALIAPORA Schlüter, 1889

Type species. Alveolites battersbyi Milne-Edwards and Haime, 1889; from the Givetian of Devonshire, England.

Subgenus CALIAPORA (CALIAPORA) Schlüter, 1889

Type species. Alveolites battersbyi Milne-Edwards and Haime, 1851; from the Givetian of Devonshire, England.

Diagnosis. Fernández-Martínez and Tourneur 1993, pp. 60–61.

Remarks. The genus *Caliapora* is composed of three subgenera: *C.* (*Caliapora*) Schlüter, *C.* (*Mariusilites*) Mironova and *C.* (*Luaciaella*) Fernández-Martínez and Tourneur. Fernández-Martínez and Tourneur (1993) regarded the nominative subgenus as the direct descendant of *Caliapora* (*Mariusilites*). The main feature of the genus treated as a whole is the presence of squamulae; their high value as taxonomical discriminator was recently evaluated by Iven *et al.* (1997). Hladil (1980) discussed the evolutionary trends of the genus *Caliapora*.

Caliapora (*C.*) *battersbyi battersbyi* (Milne-Edwards and Haime, 1851)
Figures 16E–F

* 1851 *Alveolites Battersbyi* Milne-Edwards and Haime; Milne-Edwards and Haime, p. 257.

1889 *Caliapora battersbyi* M. Edw. u. H. sp.; Schlüter, p. 95, pl. 14, figs 8–9.

v. 1939 *Caliapora battersbyi* (Milne-Edwards et Haime); Lecompte, p. 136, pl. 19, figs 1–7.

1955 *Caliapora battersbyi* (Milne-Edwards and Haime); Sokolov, pl. 33, figs 1–2.

1959 *Caliapora battersbyi* Milne-Edwards and Haime; Fontaine, p. 320.

1961   *Caliapora battersbyi* Milne-Edwards and Haime; Fontaine, p. 206.

? 1967   *Caliapora battersbyi* (M. Edwards and Haime); Tong Dzuy, p. 118, text-fig. 10a–b, pl. 23, fig. 1.

1969b   *Caliapora battersbyi* (Milne-Edwards and Haime); Stasińska, p. 773, pl. 1, figs 1–2, 4.

? 1972   *Caliapora battersbyi* (Milne-Edwards and Haime); Yanet, p. 83, text-fig. 16, pl. 26, fig. 3.

1975   *Caliapora battersbyi*; Brice, Bigey, Mistiaen, Poncet and Rohart, p. 145.

1976   *Caliapora battersbyi* (Milne-Edwards and Haime); Nowiński, p. 68, pl. 11, figs 1–2.

? 1976   *Caliapora battersbyi* (Milne-Edwards and Haime); Stasińska and Nowiński, p. 305, pl. 24, fig. 4.

1981b   *Caliapora battersbyi* (Milne Edwards and Haime); Hladil, p. 159, pl. 1, figs 3–5.

1983   *Caliapora battersbyi* (Milne-Edwards and Haime); Byra, p. 26, pl. 6, figs 12, 13, pl. 7, figs 14–16.

? 1983   *Caliapora battersbyi* (Milne-Edwards and Haime); Kulicka and Nowiński, p. 480.

? 1984b   *Caliapora battersbyi* (Milne-Edwards and Haime); Hladil, p. 252, pl. 1, fig. 2.

1985   *Caliapora battersbyi* (Milne-Edwards and Haime); Coen-Aubert, Dejonghe, Cnudde and Tourneur, p. 33, fig. 9.

1985   *Caliapora battersbyi* (Milne-Edwards and Haime); Tourneur, p. 360, text-fig. 176, pl. 33, figs 1–3.

1985   *Caliapora battersbyi* (Milne-Edwards and Haime); Birenheide, p. 64, text-fig. 16, pl. 16, fig. 1.

? 1988   *Caliapora battersbyi* (Milne-Edwards and Haime); Tong-Dzuy, Nguyen and Kromykh, p. 96, pl. 40, fig. 2, pl. 43, fig. 3.

1997   *Caliapora battersbyi* (Milne-Edwards and Haime); Iven *et al.*, text-fig. 1, pl. 1, figs 1–6.

? 1998   *Caliapora battersbyi* (Milne-Edwards and Haime); Birenheide, p. 181, pl. 8, fig. 2.

v. 2003   *Caliapora battersbyi* (Milne-Edwards and Haime); Nowiński, p. 150, pl. 97, figs 1–4.

Type material. Specimen R23432, Natural History Museum, London (see Byra 1983, p. 27).

Material. Czarnów (set A): three coralla, five thin sections (ZPAL T.25: C017–018, C034–035, C048); Laskowa (? set A): one corallum, three thin sections (ZPAL T.25: L097, L100, L101).

Description. Coralla small, up to 50 mm in largest diameter, bulbous. Corallites irregularly polygonal, isometric or elongated on cross section, 0.60–1.20 mm of the largest diameter. Walls even, thin, 0.06–0.14 mm in thickness (usually not exceeding 0.10 mm). Discontinuous, granular median line well visible. Wall microstructure diagenetically altered. Tabulae numerous, irregularly folded, complete and incomplete, spaced evenly every 0.14–0.34 mm, they are often attached to squamulae. Squamulae

numerous, 0.16–0.20 mm in length, rarely up to 0.30 mm, spaced every 0.50–0.80 mm. They are pointing up at angle 45–90 degrees. Connecting pores rare, round, 0.14–0.16 mm in diameter, spaced 0.70 mm.

Discussion. The material described here is not well preserved; therefore, numerous features cannot be observed and the pore spacing is based on one measurement. Nonetheless, it does fall within all the biometrical parameters of the neotype of *C. battersbyi* (see Byra 1983), redescribed by Tourneur *et al.* (in prep.; B. Mistiaen, pers. comm. 2007).

Occurrence. Kostomłoty Transitional Zone (Early–Middle Givetian), Montagne Noire, Boulonnais (France), Moravia, England, Vestfalia, western slopes of Ural Mts., Afghanistan and Viet-Nam (Givetian).

Caliapora (C.) battersbyi minor Nowiński, 1992

*v. 1992   *Caliapora battersbyi minor* subsp. n.; Nowiński, p. 198, 206, fig. 10.

v. 2003   *Caliapora battersbyi minor* Nowiński; Nowiński, p. 151, pl. 97, fig. 5, pl. 98, fig. 1.

Type material. Specimen ZPAL T.18-4/20 (thin sections ZPAL T.25 C043–044), Institute of Paleobiology PAS, Warsaw (Nowiński 1992, text-fig. 10 A–B).

Material. Czarnów (set A): Two coralla, three thin sections (ZPAL T.25: C043–044, C046).

Description. Nowiński (1992, p. 206) fully described this species.

Remarks. This subspecies of *C. battersbyi* was established on the basis of three poorly preserved specimens. It occurs in the same strata as the nominate subspecies; however, these subspecies differ visibly. New material would be necessary to decide whether there is a continuous distribution of features between both subspecies. At this stage and because of the absence of new material, this taxon is treated as valid. Nowiński (1992) provided good illustrations of the subspecies.

Occurrence. Kostomłoty Transitional Zone (Early Givetian), probably also Siewierz, Silesia (Givetian?).

Caliapora (C.) venusta Yanet, 1972
Figures 16G–H

*. 1972   *Caliapora venusta* sp. nov; Yanet, p. 84, pl. 27, fig. 1a–b.

1981b *Caliapora venusta* Yanet; Hladil, p. 160, pl. 4, fig. 1.

v. 1992 *Caliapora venusta* Yanet; Nowiński, p. 198, fig. 12A–B.

v. 2003 *Caliapora venusta* Yanet; Nowiński, p. 152, pl. 99, fig. 2.

Type material. Specimen 92 and/or 93/524, Institute of Geology and Geochemistry Ural Branch of RAS, Ekaterinburg, Russia (Yanet, pl. 27, fig. 1).

Material. Czarnów (set A): two coralla, four thin sections (ZPAL T.25: C038–039, 045, 047).

Description. Coralla columnar or massive, small, up to 30 mm of the largest diameter. Corallites long, straight, in axial part of coralla straight, in peripheral parts bent towards surface. In cross sections, they are polygonal, irregularly polygonal, with rounded lumina. The internal diameter (lumen diameter) of corallites varies from 0.50 to 0.82 mm, most often 0.62 to 0.80 mm. Walls even, thin, with median line often visible. The double wall thickness varies between 0.06 and 0.18 mm, most often 0.08–0.14 mm. Tabulae complete, concave, inclined, and rarely flat or folded, numerous, spaced regularly 0.30–0.70 mm, most often 0.40–0.50 mm. They are often attached to squamulae. Connecting pores uncommon, round or slightly oval, 0.10–0.18 mm of the largest diameter. Squamulae occur randomly, in certain places numerous, in others less frequent. They are long, sharp, often strongly bent towards the distal side of corallum.

Discussion. The coralla described above strongly resemble the specimens described by Yanet with this exception that the specimens described here have slightly larger maximal diameters (in the original material up to 0.70 mm of the largest diameter). All other biometrical values seem to be the same.

Occurrence. Kostomłoty Transitional Zone (Early Givetian), Western slopes of Ural Mts., Moravia (Givetian).

Subfamily NATALOPHYLLINAE Sokolov, 1950

Genus NATALOPHYLLUM Radugin, 1938

Type species. N. giveticum Radugin from the Givetian of Kuznetsk Basin.

Remarks. This genus consists of (probably) six species, ranging from Early to Late Devonian (Hill 1981, Hladil 1987).

Natalophyllum cf. *giveticum* Radugin, 1938
Figure 20D–E

*? 1938 *Natalophyllum giveticum* n. sp. Radugin, p. 79, pl. 2, figs 9, 10; pl. 5, fig. 5 (non 6).

non 1938 *Natalophyllum giveticum* var. *elegantula* nov. var. Radugin, p. 82, pl. 2, fig. 13, 21; pl. 5, fig. 6.

? 1959 *Natalophyllum giveticum* Radugin; Dubatolov p. 184, pl. 58, figs 3–5; pl. 59 fig. 1.

? 1964 *Natalophyllum giveticum* Radugin; Tchudinova, p. 58, pl. 28, figs 1–3.

? 1969b *Natalophyllum giveticum* Radugin; Stasińska, p. 775, pl. 2, figs 1–2.

? 1976 *Natalophyllum giveticum* Radugin; Nowiński, p. 71, pl. 13, fig. 3; pl. 14, fig. 1; pl. 15, fig. 3.

? 1986 *Natalophyllum giveticum* Radugin; Nowiński and Prejbisz, p. 242.

v. 2003 *Natalophyllum giveticum* Radugin; Nowiński p. 156, pl. 105, fig. 1.

Type material. Specimen 43 (coll. K. Radugin), Tomsk State University, Tomsk, Russia (Radugin 1938, pl. 5, fig. 5).

Material. Czarnów (set A): two coralla, four thin sections (ZPAL T.25 C052–055).

Description. Coralla branching. Branches moderately thick, attaining 12–14 mm in diameter. Axial zone of branch clearly marked, occupying about half of the branch diameter. Corallites

---

FIG. 20. A, *Scolipora* sp. Czarnów, Holy Cross Mountains, Poland; set A, Givetian. Specimen ZPAL T.25 C 023, transverse section. B, *Scolipora* sp. Czarnów, Holy Cross Mountains, Poland; set A, Givetian. Specimen ZPAL T.25 C 016, longitudinal section. C, *Scolipora* sp. Jaźwica Quarry, Holy Cross Mountains, Poland; set K, Early Frasnian. Specimen ZPAL T.25 JAZ 021, external view. D–E, *Natalophyllum* cf. *giveticum* Radugin, 1938. Czarnów, Holy Cross Mountains, Poland; set A, Givetian. Specimen ZPAL T.25 C052–053. D, transverse section. E, longitudinal section. F–H, *Coenites* aff. *variabilis* Sokolov, 1955. Laskowa Góra Quarry, Holy Cross Mountains, Poland; set A, Laskowa Beds, Middle Givetian. Specimen ZPAL T.25 L013–014. F, transverse section. G, longitudinal section. H, longitudinal section, detail of G showing algae(?). I, *Coenites* aff. *variabilis* Sokolov, 1955. Laskowa Góra Quarry, Holy Cross Mountains, Poland; set A, Laskowa Beds, Middle Givetian. Specimen ZPAL T.25 L039, longitudinal section. J, *Platyaxum escharoides* (Steininger, 1833). Laskowa Góra Quarry, Holy Cross Mountains, Poland; rubble of set A(?), Laskowa Beds, Middle Givetian. Specimen ZPAL T.25 LAS R031, external view. K–L, *Platyaxum escharoides* (Steininger, 1833). Laskowa Góra Quarry, Holy Cross Mountains, Poland; set A, Laskowa Beds, Middle Givetian. Specimen ZPAL T.25 LAS R050. K, transverse section. L, longitudinal section. M–N, *Platyaxum escharoides* (Steininger, 1833). Laskowa Góra Quarry, Holy Cross Mountains, Poland; set A, Laskowa Beds, Middle Givetian. Specimen ZPAL T.25 LAS 039. M, transverse section. N, longitudinal section. O–P, *Platyaxum clathratum minus* (Stasińska, 1958). Laskowa Góra Quarry, Holy Cross Mountains, Poland; set A, Laskowa Beds, Middle Givetian. Specimen ZPAL T.25 LAS 039. O, longitudinal section. P, transverse section. Scale bars represent 1 mm.

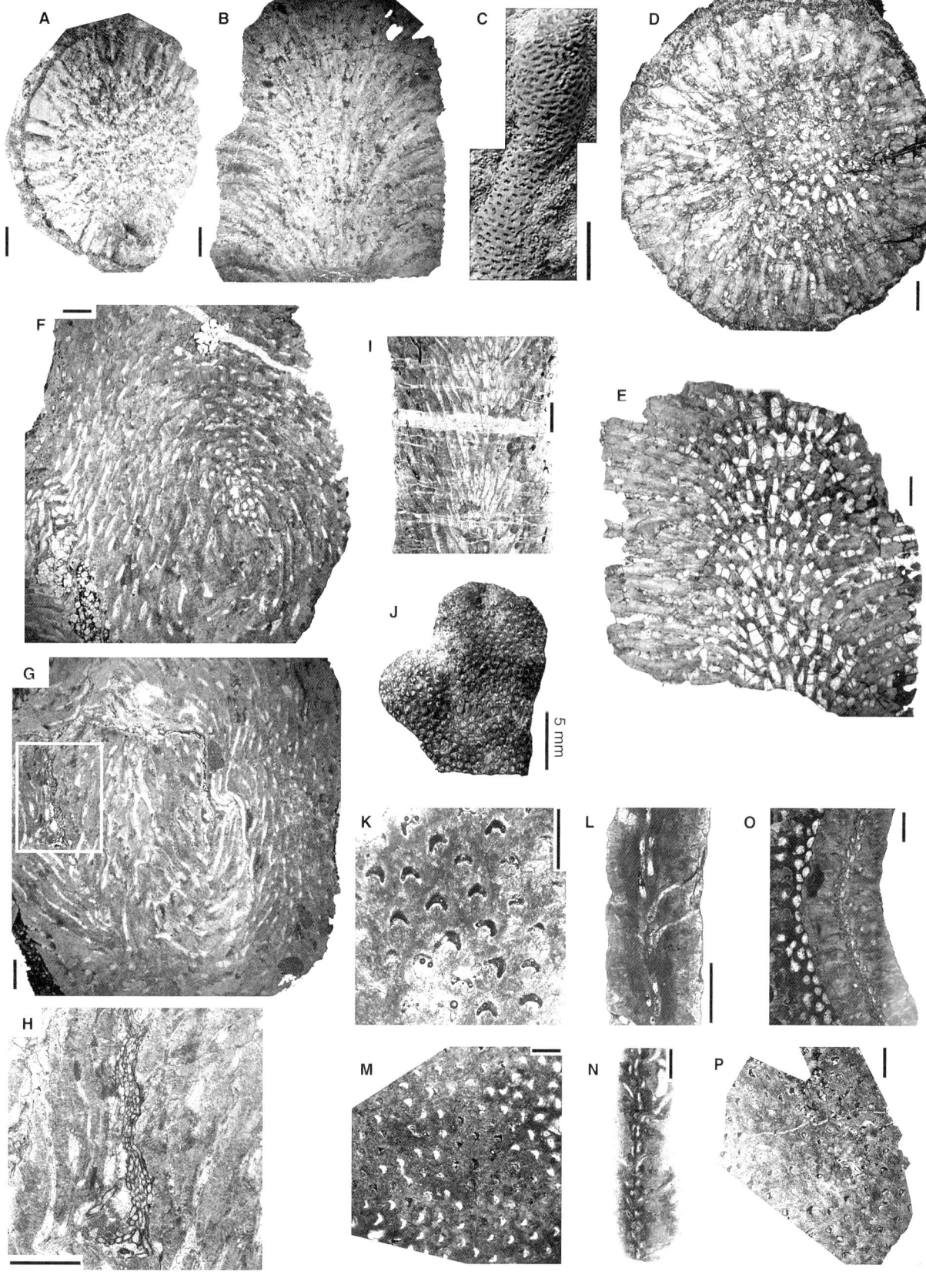

in the axial zone polygonal, 0.50–0.66 mm in diameter (max. 0.70 mm). Walls in the axial zone moderately thick, 0.04–0.18 mm, thickening evenly towards the peripheral zone. Median line 0.02–0.04 mm thick, of granular structure, in the peripheral zone becoming more even. The wall thickness in the peripheral zone probably not exceeding 0.36 mm (see remarks). Connecting pores not numerous, round, with large variation of diameter (0.1–0.2 mm), sometimes closed by poral plate. Tabulae numerous, 0.02–0.06 mm thick, flat, concave, convex, inclined; sometimes incomplete, spaced chaotically.

Remarks. The above-described specimen appears to be very similar to the species of Radugin; however, the quality of illustrations in his paper (Radugin 1938) is very poor, and in the present author's opinion, the type material needs to be revised. As not having tangential section, the wall thickness and corallite diameters in peripheral zone are taken on longitudinal section (instead of tangential). Therefore, they should be treated as approximative.

Occurrence. Northern part of Kielce Region (Early Givetian), Pomerania, Kuznetsk Basin (Givetian).

Genus SCOLIOPORA Lang, Smith and Thomas, 1940

Type species. *Scoliopora denticulata* (Milne-Edwards and Haime, 1851) from the Givetian of Belgium.

Remarks. According to Nowiński (1976), most of the representatives of genus *Scoliopora* have invisible median suture, septal spines frequently occurring also in the axial zone. Alkhovik (1985) classified this genus within Alveolitidae; the present author follows the classification of Hill (1981).

Scoliopora sp. A
Figures 20A–C

vp 2003 *Scoliopora denticulata* (Milne-Edwards and Haime); Nowiński, p. 155, pl. 102, pl. 3, pl. 103, figs 3–4.

Material. Czarnów (set A): Thin sections of four coralla (ZPAL T.25: C014–C016, C019–024, C040–43); Jaźwica (set ?K): one corallum, two thin sections (ZPAL T.25: JAZ 021.1–2); Kadzielnia (set A): two coralla, two thin sections (ZPAL T.25: Sta Kd 001–002); Psie Górki: Probably five coralla, five thin sections (ZPAL T.25: Sta PG 103–107); Wietrznia: two coralla, two thin sections (ZPAL T.25: Sta W 112–113).

Description. Coralla small, branching. Branches oval and round in cross section, rarely feebly irregular, 8–25 mm of the largest diameter. The axial zone of coralla well distinct, occupying about ½ or less of the branch diameter. Corallites in the axial parts of branches irregularly polygonal, 0.46–0.80 mm of the largest diagonal. The corallite lumina have rounded corners. Walls thin, with visible granular, uneven median line. In the peripheral zone corallites become more alveolitid, thick-walled; they open to the surface at the angle close to right. Calyces typically 'scolioporid', with one spine on one lip and two spines on the other (Fig. 20C). Tabulae rare, inclined, folded, horizontal, spaced 0.40–0.60 mm. Connecting pores about 0.10 mm in diameter, rare. Septal spines poorly preserved, short, conical.

Remarks. Nowiński (1992, 2003) mentioned this material as *S. denticulata*. Very poor preservation of the material (difficulties in basic measurements, unclear edges of septal elements) does not allow to precise determination, and it has been therefore described in open nomenclature.

Occurrence. Kostomłoty Transitional Zone (Givetian), Northern and Southern Kielce Subregions (Early Frasnian).

Scoliopora? sp. B
Figure 21A–F

v p 1992. *Striatopora* aff. *tenuis* Lecompte; Nowiński, p. 190.

Material. Kowala Railroad Section (set A and rubble): About 150 fragments of coralla, 156 thin sections (ZPAL T.25: KOW A.006.1, KOW 9.001.1, 5; KOW.R: 027.1–2; 028.1–3; 030.1–2; 031.1–2; 032.1–2; 033.1–2; 034.1–2; 035.1–2; 037.1–2; 062.1–3; 063.1–3; 064.1–3; 065.1–4; 066.1–3; 067.1–3; 068.1–3; 069.1–3; 070.1–3; 071.1–3; K007–008, K023–026, K039–045, K064–067, K076–082, K086–088, K096, K098, K106, K111–112, K122, K129–130, K158–161, K172–173, K175–183, K185–187, K189–190, K193–196, K198–200, K217–218, K227–237, K244–247, K252–256, K261–262, K264, K266–268, K283–288, K323–328, K341–342, KB001–002, KOW 362–363, KOW RO.01); Sitkówka–Kostrzewa: about fifty fragments of coralla, 19 thin sections (ZPAL T.25: SIT 3.001; ST 001–014, STA 001–004).

Description. Coralla very small. Branches slender, usually 4–5 mm in diameter, branching dichotomously. Calyces shallow, polygonal (often subrhomboidal), elongated, with rounded edges. The axial zone occupies less than half of the branch diameter. Corallites long, prismatic, reaching the surface at nearly right angle. Corallites in the axial zone polygonal with strongly rounded corners, with the diameter 0.28–0.44 mm in the axial zone, reaching 0.50 mm near the corallum surface. Walls generally thin, but variable in axial zone, thickening strongly in the peripheral zone. In the axial zone, the single wall thickness varies between 0.02 and 0.10 mm, reaching 0.90 mm near the surface of corallum. Median line blurry, poorly visible only in the axial zone, often invisible. Tabulae flat or inclined, very rare. Connecting pores very rare, 0.16 mm in diameter. Septal spines very rare, conical, in the peripheral zone occurring as *Hauptdorn*.

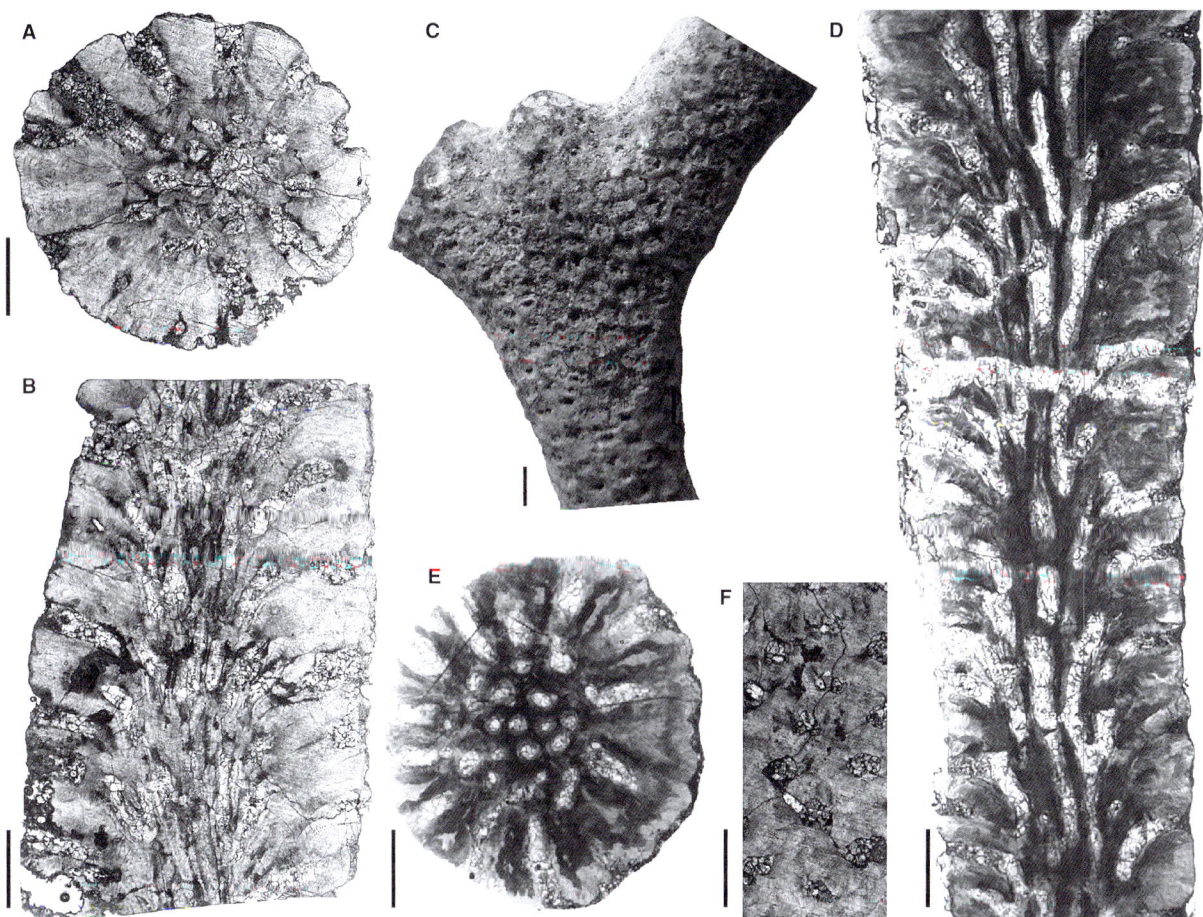

FIG. 21. A–F, *Scoliopora*? sp. B, Kowala Railroad Section, Holy Cross Mountains, Poland, rubble of set A, Early Frasnian. Specimen ZPAL T.25 KOW R.032.1–2. A, transverse section. B, longitudinal section. Specimen ZPAL T.25 KOW RO.01. C, external view of branching corallum. Specimen ZPAL T.25 KOW R.063–1.3. D, longitudinal section. E, transverse section. F, tangential section. Scale bars represent 1 mm.

Discussion. The above-described species has been temporarily assigned to genus *Scoliopora*, on the basis of shape of calyces, arrangement of corallites and occurrence of *Hauptdorn* (Fig. 6F). It resembles, however, representants of *Coenites*, from which it differs by the presence of connecting pores in the late astogenic stages. The specimens described here may represent a new genus and species; however, owing to very poor preservation of the material, the new taxa were not formally established.

Occurrence. Southern Kielce Region (Early Frasnian).

Family COENITIDAE Sardeson, 1896

Type genus. Coenites Eichwald, 1829.

Discussion. The meanders of coenitid systematics are the effect of several causes. First of all, the genus *Coenites* has

been treated as 'collective genus' for all branching and laminated coralla (even after the revision of the type species by Stumm 1960). Secondly, Spriesterbach (1935) classified his genus *Roseoporella* in hydrozoans; his paper dealt mainly with brachiopods and bivalves and was unknown for most of coral researchers. Sokolov (1952) created a new genus, *Planocoenites* (= *Placocoenites*), on the basis of *Coenites orientalis* Eichwald. The only picture of *C. orientalis* I had was that by Eichwald 1860 (pl. 20, which is not the same as mentioned by Sokolov i.e. Eichwald 1861, pl. 6, fig. 10). The specimen figured by Eichwald (1860, pl. 20) does not belong to *Coenites* at all and is very different from *Coenites juniperinus* Eichwald. The species usually assigned to genus *Planocoenites* as *Planocoenites selwinii* Nicholson (see Sokolov 1962*a*, *b*, fig. 36) probably belongs to the genus *Roseoporella* Spriesterbach, 1935.

Hill (1981) placed *Roseoporella* into synonymy of the alveolitid genus *Kitakamiia* Sugiyama, 1940; these two

TABLE 25. Comparison of three genera of the family Coenitidae, recognized by Hill (1981).

	Shape of corallum	Shape of corallites	Presence of spines	Tabulae	Corallite wall
Coenites	Slender branches	Prismatic	2 septal ridges on lower lip, 1 on upper lip	Thin, complete, horizontal, inclined	In axial parts thin, thickening
Planocoenites	Laminated	–	Septa poorly developed	Few	–
Platyaxum	Erect, flattened palmate	Subcylindrical	Not well recognized in type material	Not well recognized in type material	Thin, thickening

genera are, however, clearly distinct. May (1993a) stated that the name *Planocoenites* is junior subjective synonym of *Roseoporella* and changed the rank to subgenus of *Platyaxum* Davis (see also Brühl 1999). Following this, he distinguished the following subgenera of *Platyaxum*: *P. (Platyaxum)* Davis, 1887 and *P. (Roseoporella)* Spriesterbach, 1935; in addition, he included *P. (Egosiella)* Dubatolov *in* Sokolov, 1955 and *P. (Microalveolites)* Leleshus, 1972. In the opinion of May 1993a, *P. (Roseoporella)* is a senior subjective synonym of *Planocoenites*. In the present author's opinion, it is likely that *P. (Roseoporella) sensu* May 1993a is a senior synonym of *Planocoenites*, especially that the specimens illustrated by Spriesterbach (pl. 43, figs 3–4) show the characters typical of Coenitidae rather than those of Alveolitidae with very rare connecting pores and rare tabulae. *Microalveolites* Leleshus is most probably a synonym of *Crassialveolites* (see Hill 1981, p. F591). Hladil (1989b, p. 226) described the subgenus *Coenites (Levisicoenites)*, which is characterized by narrow calyces, stabilized thickness of the wall and compact sclerenchyme. According to Hill (1981), three genera are included to the family Coenitidae, namely *Coenites* Eichwald, *Planocoenites* Sokolov and *Platyaxum* Davis. The comparison between these genera is given in Table 25 (data after Hill 1981, pp. F600–F602).

The features of *Roseoporella* and *Egosiella* such as different corallum habit support the idea to regard these as separate genera; the systematic order within the family is shown in Table 26.

TABLE 26. Diagnostic features of genera belonging to family Coenitidae.

Genus	Diagnostic features
Coenites Eichwald	Coralla branching, erect, calyces slit, waved
Egosiella Dubatolov	Coralla branching, procumbent, calyces pocket-like
Platyaxum Davis	Coralla erect, calyces opening on both sides of corallum
Roseoporella Spriesterbach	Coralla procumbent, calyces opening on one side of corallum

Technical note. As already observed by Lecompte (1939, p. 70) sectioning of coenitids is very confusing, as the corallites are sinusoidal and meandering; Lecompte's remark concerned *Roseoporella gradata*, but it is valid for all members of the family. Therefore, in the present study, specimens of *Platyaxum* and *Roseoporella* are observed in the sections longitudinal to the growth and tangential to the surface as Lecompte also did. In several cases, sections perpendicular to the growth axis of corallum were prepared, but one must keep in mind that these are not perpendicular to the axes of the corallites.

Genus COENITES Eichwald, 1829

Type species. C. juniperinus Eichwald, 1829 from the Silurian of Lithuania.

Remarks. The genus *Coenites* is very common in Silurian and Devonian reefal environments. Stumm (1960) revised the type species stated that it is composed exclusively of branching corals. As a result, the species described by Stasińska (1958; *Coenites* in her paper) are here assigned to the closely related genera *Platyaxum* and *Roseoporella*.

Coenites aff. *variabilis* Sokolov, 1952
Figure 20F–I

v. 1992 *Coenites laminosa* Gürich; Nowiński, p. 198, fig. 12C–D.
v. 2003 *Coenites laminosa* Gürich; Nowiński, p. 152, pl. 102, fig. 2, non pl. 100, fig. 4, pl. 101, fig. 1.

Material. Laskowa Góra Quarry (probably set A): 12 specimens, 30 thin sections (ZPAL T.25: LAS 110.1–3, LAS 111.1–2, L011–014, L026–027, L039–040, L045–046, L049–053, L055–056, L075–076, L079–080, LG003–007).

Description. Coralla branching, round or oval in cross section, measuring 7–14 mm of the largest diameter. In the axial zone, corallites rounded-polygonal, their lumen diameter varies 0.12–0.24 mm, single wall thickness 0.04–0.08 mm, with visible, but blurry median line. The axial zone of branches is very narrow; it occupies 10–20 per cent of the branch diameter. Corallites twisted, they reach the surface at acute angle close to 30 degrees.

Calyces with one wall flat, the other one curved, the ratio of width/height of calyx is about 1:3. The calyx length varies between 0.32 and 0.50 mm and their height is difficult to measure, as there is a gradual passage from calyx interior to the visor-like structure (small and not always present) of neighbouring individual. Tabulae flat or concave, very rare, neither septal spines nor connecting pores not observed.

Remarks. Nowiński (1992, 2003) previously described the present material and referred the specimens to *Coenites laminosa* Gürich. According to the description by Stasińska (1958), it can be stated that *C. laminosa* owing to the 'form of stratified plates' (Stasińska 1958, p. 219) and calyces opening on both surfaces (Stasińska 1958, pl. 30, fig. 2) belongs clearly to *Platyaxum*. The corals described here, being in part material of Nowiński (1992), belong undoubtedly to the genus *Coenites* owing to the branching habit of coralla.

The species investigated here shows several similarities to *Coenites variabilis* Sokolov, 1952 from the Frasnian of the Volgograd Region. Both species have similar size of calyces and proportion of axial zone to peripheral zone. They differ in the branch diameter and development of tabulae, both of which may be controlled ecologically. One of the specimens has algae-like organisms incorporated between the walls (Fig. 20H).

Occurrence. Kostomłoty Transitional Zone (Early–Middle Givetian).

Genus PLATYAXUM Davis, 1887

Type species. Platyaxum turgidum Davis, 1887 (= *P. undosum* Davis, see Stumm 1965) from the Middle Devonian of the Falls of the Ohio, USA.

Remarks. Species of *Platyaxum* have been assigned to *Coenites* (e.g. Stasińska 1958) or to *Placocoenites* (Sokolov 1952, = *Planocoenites* Sokolov), which is a junior subjective synonym of *Platyaxum*. *Coenites* differs from *Platyaxum* in the habit of corallum, which is palmate, but erect in *Platyaxum*, and forms slender branches in *Coenites*. Another closely related genus is *Roseoporella* (see below), which also has palmate coralla, but laying on the substratum (prostrate); thus, calyces open only on one (upper) surface, while *Platyaxum* has calyces on both sides of corallum as seen in *P. escharoides* (Steininger).

Platyaxum clathratum minus (Stasińska, 1958)
Figure 20O–P

   * v. 1958 *Coenites clathratus minor* ssp. n. Stasińska,
        p. 216, pl. 27, figs 1–3, pl. 28, fig. 4.

     v. 2003 *Coenites clathratus minor* Stasińska; Nowiński,
        p. 152, pl. 100, figs 1–3.

Type material. Specimen ZPAL T.2/462, coll. A. Stasińska, Institute of Paleobiology, Warsaw, Poland (Stasińska 1958, pl. 28, fig. 4).

Material. Laskowa: three coralla, eight thin sections (ZPAL T.25: L2.030.1–2, L2.032, LAS R.050.1–4).

Description. Coralla small, palmate, cambered, 2–3 mm in thickness. Calyces semicircular and crescentic, lumen 0.12–0.20 × 0.20–0.32 mm. Corallites in the axial zone oval and round in cross section, 0.12–0.20 mm of the largest diameter. They are meandering and reach the surface at acute angle, about 60 degrees. Walls very thick, 0.36–0.50 mm of double wall thickness. Median line in the central zone of coralla visible, in the more distal zones lacking. Tabulae very rare, septal spines absent, connecting pores not observed on material. Visor present, but often not preserved.

Discussion. The subspecies described by Stasińska (1958) is assigned to *Platyaxum* on the basis of the flattened, palmate, cambered habit of the coralla, which are branching in *Coenites*. Stasińska (1958, p. 216) used the term '*polypiers ramifiés*', but in this case, the term was used in a broader context and also covered the 'platyaxum type' of corallum.

The Givetian material differs from the originally described material derived from the Emsian–Eifelian at Wydryszów, Łysogóry Region of the Holy Cross Mountains by having slightly larger minimal lumen diameters, which may be due to a different angle of sectioning. Because of scarcity and poor preservation of the material, the statistical study on intracolonial variation has not been executed.

Occurrence. Łysogóry Region (Emsian–Eifelian), Kostomłoty Transitional Zone (?Early–?Middle Givetian).

Platyaxum escharoides (Steininger, 1849)
Figure 20J–N

   * 1849 *Limaria escharoides* Steininger, p. 11.
     1889 *Coenites escharoides* Stein. sp.; Schlüter, p. 126,
        pl. 5, fig. 12–13.
     1908 *Coenites escharoides* (Steininger); Reed, p. 25,
        pl. 4, fig. 5–6.
     1939 *Coenites escharoides* (Steininger); Lecompte, p. 65,
        pl. 11, fig. 5–7.
     1959 *Placocoenites escharoides* (Steininger); Dubatolov,
        p. 174, pl. 55, fig. 4.
 ? 1969*b* *Placocoenites escharoides* (Steininger); Stasińska,
        p. 774.
 ? 1980 *Placocoenites* cf. *escharoides* (Steininger); Iven,
        p. 158, pl. 11, fig. 5–7.
     1985 *Platyaxum* (*Platyaxum*) *escharoides* (Steininger);
        Birenheide, p. 86, pl. 30, fig. 1.

1987   *Placocoenites escharoides* (Steininger); Sarnecka, p. 135, pl. 5, fig. 3.

1993a   *Platyaxum (Platyaxum) escharoides* (Steininger); May, p. 171, pl. 9, fig. 3–4.

Type material. Unknown (see May 1993a, p. 171).

Material. Laskowa Quarry: 20 incomplete coralla (LAS R.006, LAS R.030–038, LAS R.040, LAS R. 041, LAS R.045, LAS R050–056), 11 thin sections (LAS R.006.1–2, LAS R. 050.1–2, LAS R. 051.1–2, LAS R.053.1–2, LAS R.054, LAS R.055.1–2).

Description. Coralla small, palmate, cambered, thin, 1.5–3.0 mm in thickness, with calyces opening on both sides. Calyces horse-shoe shaped, measuring 0.12–0.18 × 0.30–0.34 mm. Corallites curved, very small in cross section, in the proximal parts thin-walled, oval, with walls thickening significantly towards surface, where the corallites gradually change shape into crescentic. They reach the surface of corallum at the acute angle, close to 30 degrees. Median line is not visible. Tabulae and pores are very rare. Septal spines very rare, visor absent.

Measurements. The biometrical values are given in Table 27.

Remarks. This very common species is relatively difficult to study. Its lamellar, thin, cambered coralla do not allow obtaining large thin sections, and the neighbouring corallites are moreover always cut on different levels, giving different wall thicknesses and corallite diameters. Additionally, the material with good 'external look' is usually poorly preserved.

The intracolonial variety of this species remains low, especially when compared to closely related *Roseoporella*. The largest diameter of lumen has low variation coefficient, while its counterpart, smallest lumen diameter changes slightly more. The variation of double wall thickness remains very low.

Technical comment. Measurements of calyces were taken on the surface of corallum, while lumen diameters on the section tangential to the corallum surface.

TABLE 27. Intracolonial variation in *Platyaxum escharoides* (Steininger).

	Max LD	Min LD	DWT
Specimen ZPAL T.25 LAS 006, Laskowa Góra Quarry, Givetian			
Mean	0.309	0.149	0.577
Standard deviation	0.0381	0.0261	0.0675
N	30	30	30
Minimal value	0.22	0.10	0.46
Maximal value	0.40	0.20	0.70
V	0.124	0.175	0.117

Measurements in millimetres. For abbreviations, see Table 1.

Occurrence. Kostomłoty Transitional Zone (?Early–?Middle Givetian), Rheinisches Schiefergebirge, Kuznetsk Basin (Givetian); Padaupkin (Burma, ?Middle Devonian).

Genus ROSEOPORELLA Spriesterbach, 1935

Type species. R. rhenana Spriesterbach, 1935 from the Middle Devonian of Rheinisch Slate Mountains.

Remarks. Representatives of this genus have been for a long time described under generic names *Alveolites*, *Planocoenites* (= *Placocoenites*) and *Coenites* (e.g. Stasińska 1958; Mironova 1974; Kamiya and Niko 1998; Niko 2003). From *Alveolites*, it differs principally by very scarce connecting pores, also by laminated corallum, shape of corallites, which most often are crescentic in the discussed genus, but alveolitoid-shaped are also present, and less numerous tabulae. It differs from *Coenites* and *Platyaxum* by habit of corallum, which is prostrate and laminar in the discussed genus. One more characteristic, but not diagnostic, feature is the occurrence of 'rosebud' arrangement of corallites (see Spriesterbach 1935, pl. 42, fig. 4).

Roseoporella heuvelmansi sp. nov.
Figure 22A–E

Derivation of the name. In honour of the late Dr. Bernard Heuvelmans, for his work on aardvark dentation.

Type strata. Laskowa Góra Beds (set B), Middle Givetian.

Type locality. Laskowa Góra Quarry, Holy Cross Mountains, Poland.

Holotype. ZPAL T.25 LAS R.001.1–3 (Fig. 22A–C); one corallum, three thin sections.

Additional material. Laskowa Góra Quarry (set B): four coralla, nine thin sections (ZPAL T.25: LAS R.004.1–2, LAS R.007.1–2, LAS R.045.1–2).

Diagnosis. Coralla prostrate, small, very thin. Corallites variable in shape, most often lenticular and irregular, their lumen measuring 0.24–0.50 × 0.40–0.90 mm. Wall distally very thick, DWT varies between 0.14 and 0.70 mm (mean: 0.4). Connecting pores scarce, 0.08–0.20 mm in diameter, tabulae spaced 0.20–1.10 mm.

Description. Coralla prostrate, small and medium, very flat. The holotype is 3 mm thick and about 30 mm of the largest diameter; other specimens attain 5 mm of thickness and up to probably 35 mm. The bottom side of coralla is composed of very long, prostrate corallites; from the bottom corallum looks like a palm leaf. The corallites are curving upwards, and they reach the surface of corallum at acute angle, usually 45–60 degrees. In the initial stages (horizontal corallites, i. e. prostrate part), they are oval, with lumen measuring 0.20–0.36 × 0.30–0.60 mm and double wall thickness 0.08–0.18 mm. In the distal parts of the corallum, corallites in cross sections are lens-shaped (most often), irregular or rarely semicircular. Walls in the proximal parts of corallum thin, they show remarkable distal thickening (Fig. 22A, E). Median line invisible. Connecting pores rare, 0.08 0.20 mm in diameter, with unknown spacing. Tabulae rare, folded, irregular, inclined. Septal spines very rare, scattered, distributed irregularly.

Measurements. Biometrical values are given in Table 28.

Discussion. The specimens described above have been assigned to genus *Roseoporella* on the basis of such features as (1) prostrate, laminar coralla with 'palm leaf' pattern of lower side of coralla; (2) scarcity of connecting pores; (3) distal thickening of the wall; (4) corallites reaching the corallum surface at acute angle; this character allows to distinguish *Crassialveolites* from *Roseoporella* in these species of *Crassialveolites*, where connecting pores are scarce. The new species differs from *R. gradata* (Lecompte) by having different shape of corallites, which are of crescentic shape in the latter species, thicker wall, that is, up to 0.45 mm in *R. gradata* and finally by considerably thinner coralla (7–40 mm in *R. gradata*). *R. media* (Lecompte) has more steep angle of corallite opening (50–80 degrees) and much thicker corallum, that is, up to 15 mm in thickness in *R. media*. From all other known species of *Roseoporella*, it differs by enormously thick corallite wall and noncrescentic shape of calyces.

The discussed species shows very high intraspecific variability. Two reasons of such a situation are possible (1) high intracolonial variations (high variation coefficients within corallum) and (2) slightly varying angle of sectioning. In the discussed case, both factors are probably overlapping as: (1) usually coralla differ significantly by only one character; (2) owing to the geometry and size of coralla (very flat, 3–5 mm thick), even slight change of the angle of sectioning (about 5 degrees) can change the observed size of elements because of intersection; and (3) overall high variation coefficients (especially for the double wall thickness). The holotype has been chosen because the angle of sectioning seems to be the most correct; it well displays features of the species and the part of corallum remaining after the sectioning displays well the 'palm leaf' arrangement of corallites. The specimen ZPAL T.25 L091–092 has somewhat larger maximal lumen diameters and has been assigned to the new species with a doubt.

Occurrence. Kostomłoty Transitional Zone (Early–Middle Givetian).

Roseoporella sp. A
Figure 23A–B

Material. Laskowa Góra Quarry (set ?B): one incomplete corallum, four thin sections (ZPAL T.25 LAS R.044.1–4).

Description. Corallum prostrate, medium sized, very flat. Its thickness varies between 3 and 6 mm, the largest length of this fragment is about 50 mm. The bottom side of corallum is composed of very long, prostrate corallites with a 'palm leaf

TABLE 28. Intracolonial variation in *Roseoporella heuvelmansi* sp. nov.

	Max LD	Min LD	DWT	PD	TS
Specimen ZPAL T.25 LAS R.001, Laskowa Quarry, Givetian. Holotype					
Mean	0.570	0.323	0.442	–	–
Standard deviation	0.1233	0.0497	0.1259	–	–
N	31	31	31	4	5
Minimal value	0.3	0.24	0.14	0.10	0.46
Maximal value	0.9	0.5	0.70	0.14	1.08
V	0.216	0.154	0.284	–	–
Specimen ZPAL T.25 ZPAL L091–092, Laskowa Quarry, Givetian					
Mean	0.710	0.384	0.434	–	0.653
Standard deviation	0.1551	0.0611	0.1181	–	0.2379
N	32	32	32	3	11
Minimal value	0.40	0.22	0.20	0.06	0.20
Maximal value	1.04	0.48	0.70	0.10	1.10
V	0.218	0.159	0.272	–	0.364

Measurements in millimetres. For abbreviations, see Table 1.

TABLE 29. Intracolonial variation in *Roesoporella* sp. A.

	Max LD	Min LD	DWT	PD	TS
Specimen ZPAL T.25 LAS R.044, Laskowa Quarry, Givetian					
Mean	0.820	0.480	0.408	–	0.626
Standard deviation	0.1875	0.0908	0.1452	–	0.1745
N	30	30	30	7	30
Minimal value	0.54	0.34	0.12	0.10	0.32
Maximal value	1.32	0.70	0.78	0.20	0.96
V	0.229	0.189	0.356	–	0.279

Measurements in millimetres. For abbreviations, see Table 1.

pattern. The corallites are curving upwards, and they reach the surface of corallum at acute angle, close to 60 degrees. In the distal parts of the corallum corallites in cross sections are lens-shaped (most often), semicircular or irregular. Walls show distal thickening, while in the proximal parts of corallum they are thin. Median line visible in between several corallites. Connecting pores numerous in some places, 0.10–0.20 mm in diameter, with unknown spacing. Tabulae feebly or strongly concave, folded, irregular, inclined. Septal spines rare, short, conical, distributed irregularly.

Measurements. Table 29.

Remarks. The specimen described above resembles *R. heuvelmansi* sp. nov. by general shape of corallum, similar wall thickness, tabulae spacing and pore diameters (they are however slightly larger in the discussed species). It differs by larger size of visceral chamber (lumen) and much more frequent tabulae. The intracolonial variation in this specimen is very high, comparing to *Roseoporella heuvelmansi* sp. nov. and *Roseoporella* sp. B.

Occurrence. Kostomłoty Transitional Zone (?Middle Givetian).

Roseoporella sp. B
Figure 22F–G

Material. Laskowa Góra Quarry (sets A–?B): 12 coralla, 30 thin sections (ZPAL T.25 LAS R.002.1–3, LAS R.003.1–4, LAS R.005.1–2, LAS R.009.1–2, LAS R.013.1–3, LAS 015.1–3, LAS 025.1–6, LAS R042.1–3, L073–074, L094–095).

Description. Coralla prostrate, medium to large sized and very flat. Their thickness varies between 3 and 5 mm; the maximal diameter of corallum reaches 100 mm. The bottom side of coralla is composed of very long, prostrate corallites; from the bottom, corallum looks like a palm leaf. The corallites are curving upwards, and they reach the surface of corallum at acute angle, usually 45–60 degrees. In the initial stages (horizontal corallites, i.e. prostrate part), they are oval and round in cross section. In the distal parts of the corallum, corallites in cross sections are lens-shaped (most often), semicircular or irregular. Walls show distal thickening, while in the proximal parts of corallum, they are thin. Median line is invisible. Connecting pores rare, with unknown spacing. Tabulae concave, folded, irregular, inclined. Septal spines very rare, short, conical, distributed irregularly.

Measurements. Table 30.

Discussion. The described specimen differs from *R. heuvelmansi* sp. nov. by having much larger coralla, thinner corallite walls and more frequent tabulae; however, in specimen ZPAL T.25 L091–092 (*R. heuvelmansi*), the large lumen diameter is even larger than at *R.* sp. B. Also the coralla of *R.* sp. B. are much larger than those of *R. heuvelmansi* sp. nov. The intracolonial variation is high, comparable to *R. heuvelmansi* sp. nov., but much smaller than in *Roseoporella* sp. A.

Occurrence. Kostomłoty Transitional Zone (Givetian).

Order SYRINGOPORIDA Sokolov, 1962*a, b*

Family SYRINGOPORIDAE de Fromentel, 1861

Type genus. Syringopora Goldfuss, 1826.

FIG. 22. A–C, *Roseoporella heuvelmansi* sp. nov., holotype. Laskowa Góra Quarry, Holy Cross Mountains, Poland; set A, Laskowa Beds, Middle Givetian. Specimen ZPAL T.25 LAS R001. A, transverse section. B, longitudinal section. C, external view, proximal side of the corallum showing 'palm leaf' pattern. D–E, *Roseoporella heuvelmansi* sp. nov., Laskowa Góra Quarry, Holy Cross Mountains, Poland; set A, Laskowa Beds, Middle Givetian. Specimen ZPAL T.25 LAS R004. D, longitudinal section. E, longitudinal section. F–G, *Roseoporella* sp. B, Laskowa Góra Quarry, Holy Cross Mountains, Poland; set A, Laskowa Beds, Middle Givetian. Specimen ZPAL T.25 LAS R025. F, longitudinal section. G, longitudinal section. Scale bars represent 1 mm, if not stated otherwise.

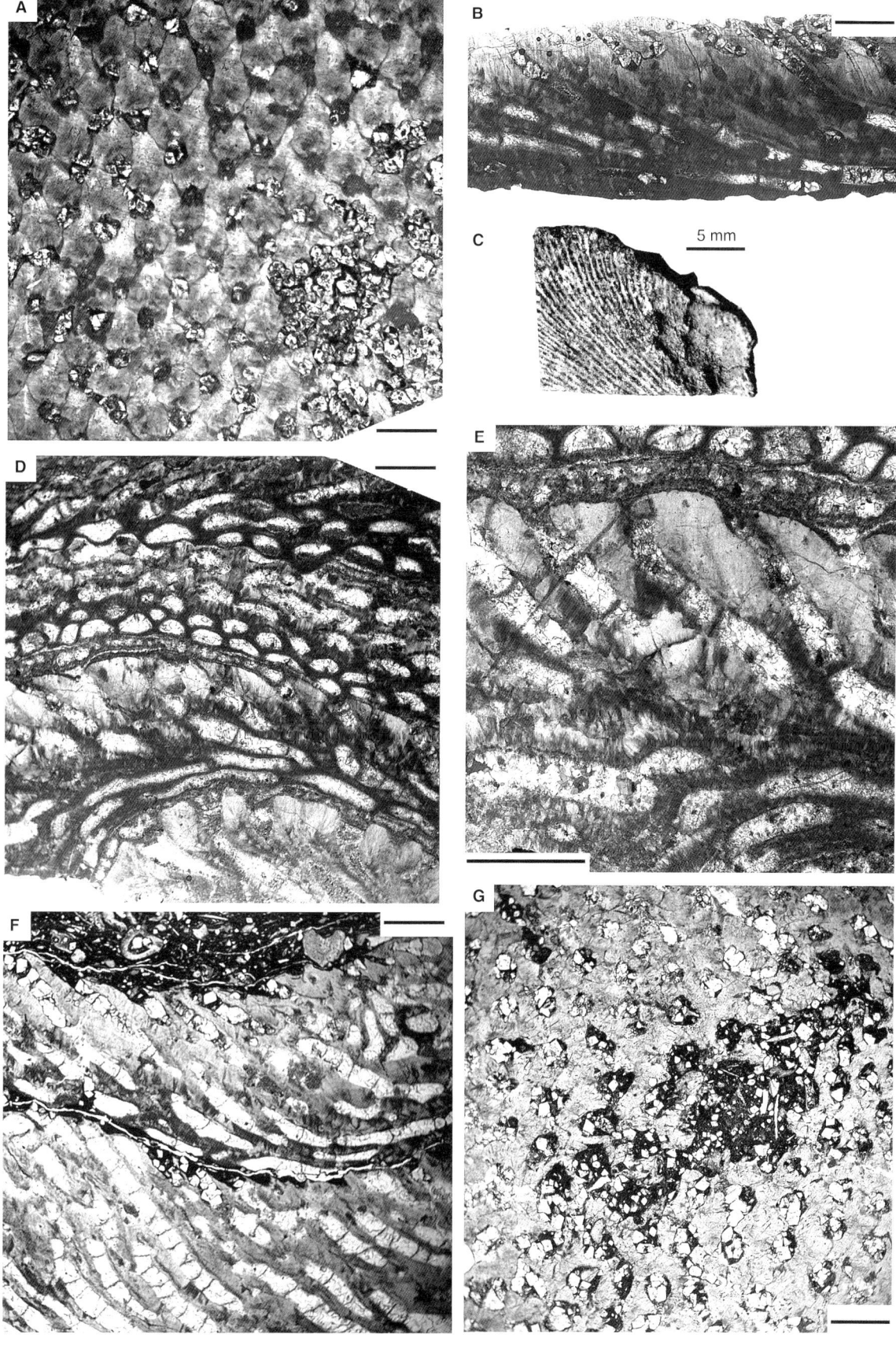

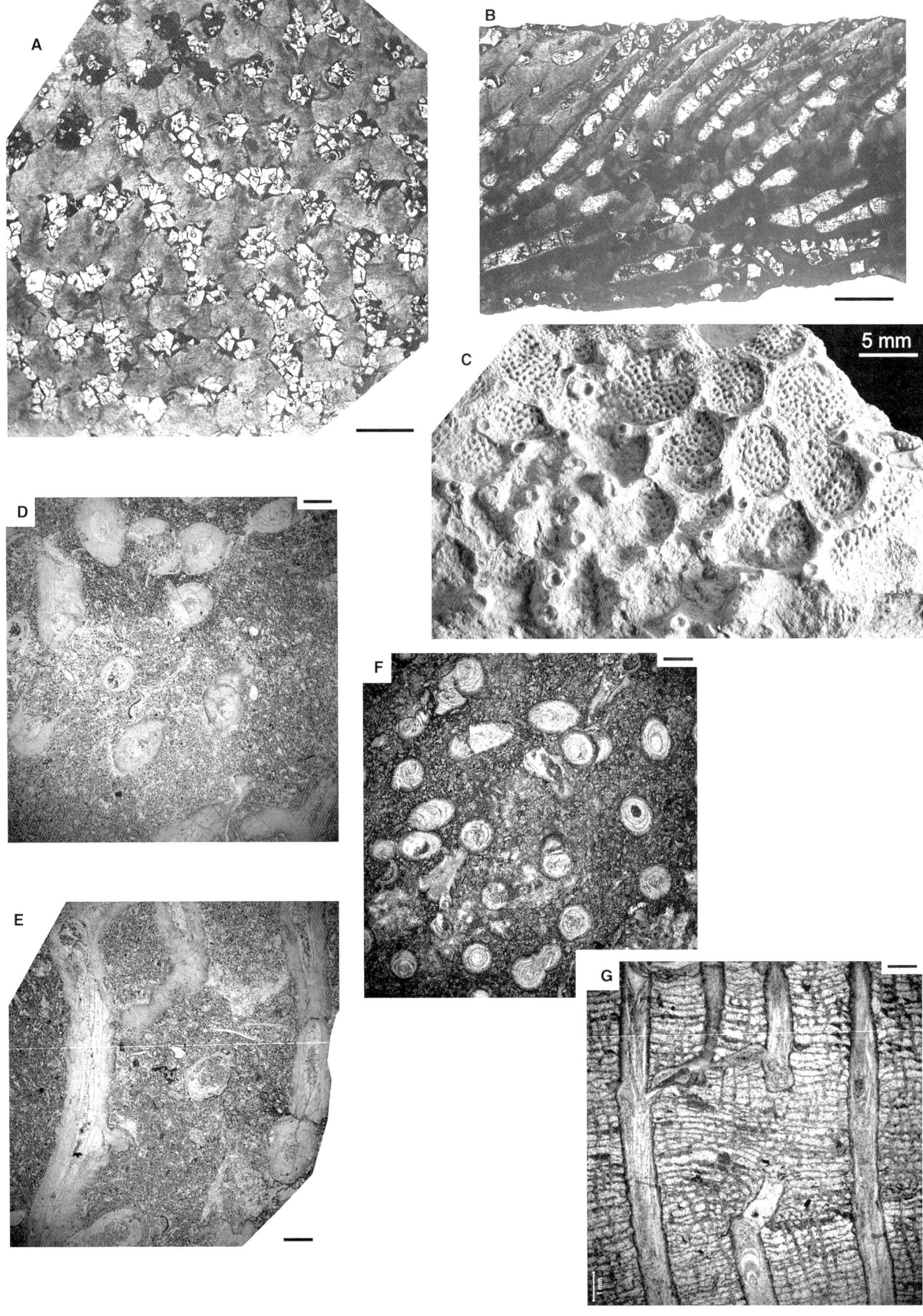

TABLE 30. Intracolonial variation in *Roseoporella* sp. B.

	Max LD	Min LD	DWT	PD	TS
Specimen ZPAL T.25 LAS R.025, Laskowa Quarry, Givetian					
Mean	0.632	0.389	0.297	–	–
Standard deviation	0.1160	0.0702	0.1035	–	–
N	31	31	31	4	5
Minimal value	0.42	0.22	0.12	0.10	0.46
Maximal value	0.86	0.56	0.54	0.14	1.08
V	0.184	0.180	0.349	–	–

Measurements in millimetres. For abbreviations, see Table 1.

Remarks. Hill (1981) included the following genera in the family Syringoporidae: *Syringopora* Goldfuss, *Cannapora* Hall, *Chia* Lin, *Enigmalites* Tchudinova, *Kueichowpora* Chi, *Oharaia* Nelson, *Pleurosiphonella* Tchudinova, *Syringoalcyon* Termier and Termier, *Syringocolumna* Stumm and *Syringoporiella* Rukhin. Tchudinova (1986) divided the family into two subfamilies, but included also representatives of the family Gorskyitidae Lin, such as *Gorskyites* Sokolov. Nowiński (1991) also referred the genera *Fuchungopora* Lin and *Roemerolites* Dubatolov (= *Armalites* Tchudinova; see Nowiński 1991, p. 49) to Gorskyitidae. The Tchudinova's subdivision of Syringoporida into two subfamilies is somewhat diffuse, and the systematics promoted by Hill (1981) is adopted in this paper but with the following changes:

1. The genus *Maksymilianites* Zapalski and Nowiński (= *Syringella* Nowiński) is considered a distinct genus, and not a junior synonymy of *Chia* Lin, as otherwise advocated by Hill (1981).

2. The genus *Armalites* Tchudinova is considered as a junior subjective synonymy of *Roemerolites* thus following Nowiński (1991, p. 49).

3. The genus *Fuchungopora* Lin is included in the family Syringoporidae, following Nowiński (1991).

4. The newly described genus *Sapounofouskilites* gen. nov. is allocated to Syringoporidae.

As a result, the family Syringoporidae is composed of the following genera: *Cannapora* Hall, *Chia* Lin, *Enigmalites* Tchudinova, *Fuchungopora* Lin, *Kueichowpora* Chi, *Maksymilianites* Zapalski and Nowiński, *Oharaia* Nelson, *Pleurosiphonella* Tchudinova, *Roemerolites* Dubatolov, *Syringoalcyon* Termier and Termier, *Syringocolumna* Stumm, *Syringopora* Goldfuss, *Syringoporiella* Rukhin, and *Sapounofouskilites* gen. nov.

Genus MAKSYMILIANITES Zapalski and Nowiński, 2005

1970 *Syringella* Nowiński, p. 540.

Type species. Syringella polonica Nowiński, 1970 (Givetian–Frasnian); from Sowie Górki, Holy Cross Mountains.

Remarks. The genus *Syringella* has been considered a subjective synonymy of *Chia* Lin (i.e. Hill 1981; Mistiaen 1988). After reinvestigating the type material, Tchudinova (1986) found that it is a valid genus based on the vesicles in the corallite wall and thick axial canal as the main features.

The original genus name *Syringella* Nowiński, 1970 (non *Syringella* Schmidt, 1868 = Porifera) has been changed to *Maksymilianites* because it was preoccupied (see Zapalski and Nowiński 2005).

Maksymilianites polonicus (Nowiński, 1970)
Figure 24A–D

* v. 1970 *Syringella polonica* sp. n. Nowiński, p. 540, text-fig. 1–3, pl. 1, fig. 1–4, pl. 2, figs 1–3.

. 1985 *Syringocystis polonica* (Nowiński); Birenheide, p. 134, text-fig. 84.

FIG. 23. A–B, *Roseoporella* sp. A, Laskowa Góra Quarry, Holy Cross Mountains, Poland; set A, Laskowa Beds, Middle Givetian. Specimen ZPAL T.25 LAS R044. A, transverse section. B, longitudinal section. C. *Aulopora slosarskii* sp. nov., holotype, encrusting *Crassialveolites oliveri* sp. nov., holotype. Kowala Quarry, Holy Cross Mountains, Poland; rubble of sets A–C, Early Frasnian. Specimen MZHT 210/D. External view. See also Figures 18A–F and 26C–D. D–E, *Adetopora*? *tikhyi* Sokolov, 1952. Biohermal complex (set C) in the Kowala Quarry, Holy Cross Mountains, Poland; Frasnian, below *disparilis* conodont zone. Specimen ZPAL T.25 KQ 001–002. D, transverse section. E, longitudinal section. F–G, *Syringoporella* sp. overgrown by Actinostromatidae indet. Jaźwica, Holy Cross Mountains, Poland; set K, Early Frasnian. Specimen ZPAL T.25 J089–090. F, longitudinal section. G, longitudinal section. Scale bars represent 1 mm, if not stated otherwise.

. 1986   *Syringella polonica* Nowiński; Tchudinova,
p. 117, text-fig. 53.

v ? 1988   *Chia polonica* (Nowiński); Mistiaen, p. 214, text-
fig. 14, pl. 26, figs 11–14.

v. 2003   *Syringella polonica* Nowiński; Nowiński, p. 158,
pl. 108, fig. 2.

v. *2005   Maksymilianites polonicus* (Nowiński); Zapalski
and Nowiński, text-fig. 1a–e.

Material. Sowie Górki Quarry: One corallum (ZPAL T. 5/1, holotype), eight thin sections (unnumbered), 85 plastic imprints.

Description. Nowiński (1970, pp. 542–543) fully described this species.

Remarks. The specimen described by Mistiaen (1988) seems to be very similar to the type specimen; however, it has smaller diameters, and the vesicles in the walls are less frequent.

Occurrence. Central Kielce Subregion (Early Frasnian), Boulonnais (?; Middle Givetian).

Genus SAPOUNOFOUSKILITES gen. nov.

Derivation of the name. From Greek *sapounofouska* meaning a bubble, because of the presence of dissepimental tissue near the wall.

Type species. Armalites minimus Nowiński, 1992; late Frasnian, the Holy Cross Mts., by monotypy.

Diagnosis. Coralla small, fascicular or locally subcerioidal. Corallites in cross section cylindrical, polygonal when in contact. Epitheca thin, walls thick and very thick, often nearly closing the lumen. Tabulae numerous, deeply infundibuliform, sometimes incomplete forming in places dissepimental structure near the wall. Connecting tubuli present when corallites spaced, otherwise pores. Axial canal (syrinx) variable, thick- or thin-walled, centrally or peripherally placed. Septal spines in rows.

Remarks. The new genus is closely related to the genera *Roemerolites*, *Chia* and *Maksymilianites*. The new genus differs from *Roemerolites* by the presence of dissepimental structure adhering to the wall, which is a feature that occurs in the two other genera. It is different from *Chia* by having very thick walls and thick-walled axial canal

and from *Maksymilianites* by having very thick walls and vesicles are absent.

The stereoplasma of the walls in *Sapounofouskilites* gen. nov. may be composed of two layers; the inner layer is composed of concentrically arranged elements, and the outer one is composed of radially arranged microcrystals (see the description of the type species). This character is not included in the diagnosis, as it does not occur equally in all corallites and it may be a diagenetical artefact.

The type species *S. minimus* displays intermediate characters between the genera *Roemeripora* and *Armalites*. It does however have its own qualitative character, which is the dissepimental structure near the corallite wall. Another unique character for this genus is the large thickness of walls in distal parts of the corallum; such thick walls do not occur in any other species of the family.

Sapounofouskilites minimus (Nowiński, 1992)
Figure 25A–K

v non 1978   *Roemerolites lublinensis* n. sp. Stasińska and
Nowiński, p. 204, pl. 17, fig. 1a–c.

v. 1992   *Roemerolites lublinensis* Stasińska and Nowiński; Nowiński, p. 188.

* v. 1992   *Armalites minimus* Nowiński; Nowiński,
p. 208, fig. 13C–D.

vp. 2003   *Roemerolites lublinensis* Stasińska and Nowiński; Nowiński, p. 130, pl. 70 fig. 4, pl. 71,
fig. 1.

v. 2003   *Armalites minimus* Nowiński; Nowiński,
p. 158, pl. 108, fig. 1.

Type material. Jaźwica Quarry (set R): one specimen, three thin sections (ZPAL T.18-14/34; T.25 J094–096, holotype = Fig. 25A–F.)

Additional material. Bolechowice, Panek Quarry: three incomplete, otherwise well-preserved coralla, ZPAL T.18–2/3 (two coralla in one specimen) and T. XVIII–1/3 (13 thin sections: BP 001–009, and four thin sections marked same as samples); a single section without specimen (STA BP 001); three ultrathin sections.

Description. Coralla discoidal, measuring about 30–40 mm in diameter. Calyces very deep, with sharp edges (Fig. 25K). Corallites loosely spaced in most parts of corallum, rarely in contact. Corallites short, tubular, round or nearly round in cross section, in few places slightly polygonal (when in contact). Walls very

---

FIG. 24. A–D, *Maksymilianites polonicus* (Nowiński, 1970). Holotype. Sowie Górki Quarry, Holy Cross Mountains, Poland. Late Givetian–Early Frasnian. Specimen ZPAL T.25 T. V/1. A, transverse section. B, transverse section. C, longitudinal section. D, longitudinal section. Arrow shows well-developed septal spine. Note also the vesicles diagnostic for the genus. E–F, *Syringopora* cf. *volkensis* Tchernychev, 1938. Jaźwica/Góra Łgawa, Holy Cross Mountains, Poland; set R, Late Frasnian. Specimen ZPAL T.25 J024–026. E, transverse section. F, longitudinal section. G–H, *Syringopora? tikhyiformis* (Stasińska and Nowiński, 1978) Jaźwica/Góra Łgawa, Holy Cross Mountains, Poland; set R, Late Frasnian. Specimen ZPAL T.25 J091–093. G, transverse section. H, longitudinal section. Arrow shows connecting tube. Scale bars represent 1 mm.

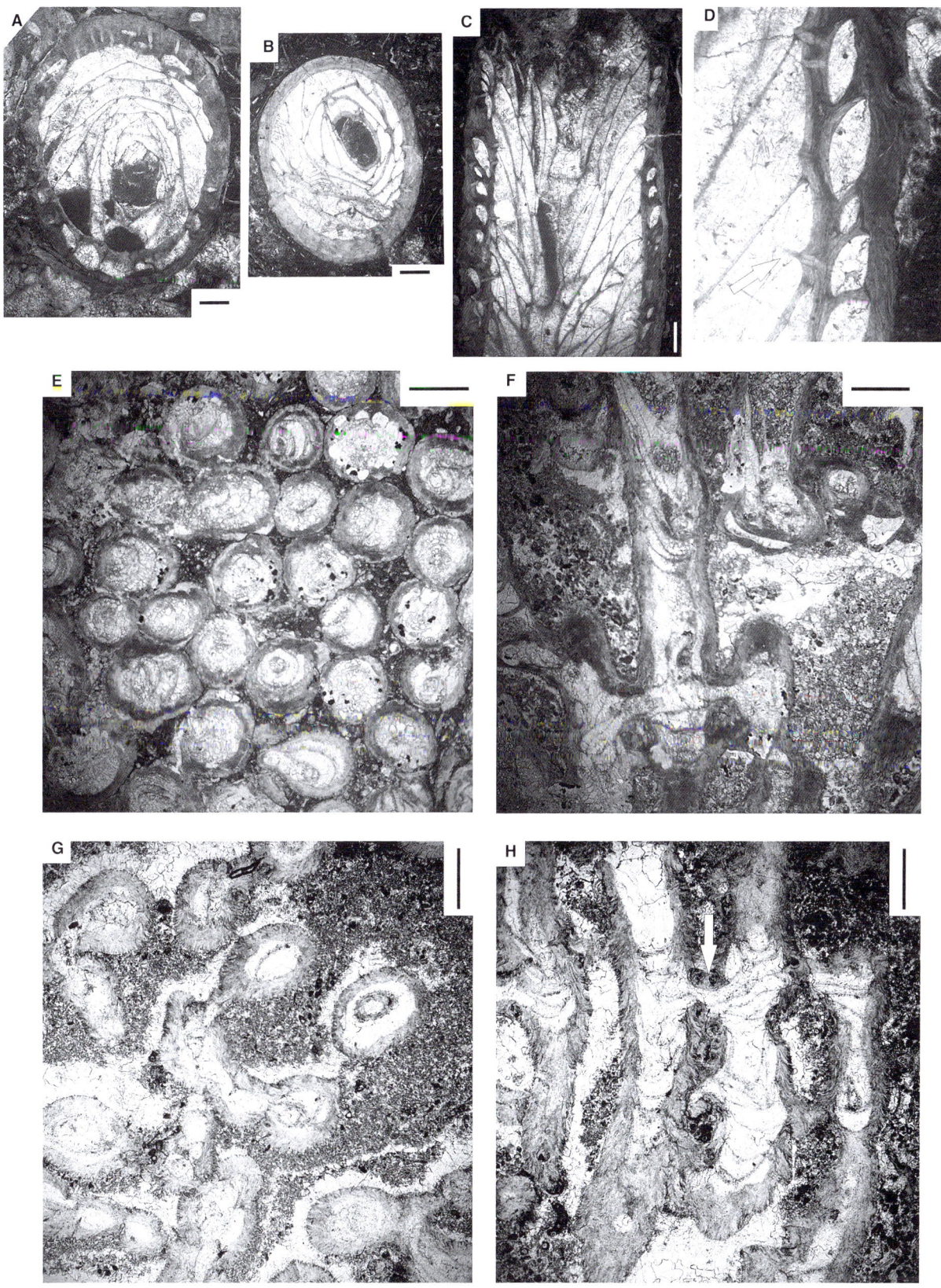

TABLE 31. Intracolonial variation in *Sapounofouskilites minimus* (Nowiński).

Character	Specimen		
	Parameter	ZPAL T.XVIII 14/34 (holotype) Jaźwica Quarry, Frasnian	TXVIII–2/3 Bolechowice, Panek Quarry Frasnian
Corallite diameter	Mean	1.307	1.369
	Standard deviation	0.1301	0.0697
	Variation coefficient	0.100	0.051
Single wall thickness	Mean	0.451	0.421
	Standard deviation	0.1116	0.0990
	Variation coefficient	0.247	0.235

Measurements in millimetres.

thick, occupying up to three-fourth of corallite diameter, uneven, with epitheca irregularly wrinkled. Connecting tubuli rare, occurring irregularly throughout the corallum. Connecting pores not common, occur only in several places, where corallites are in contact. Pore diameter ranges from 0.20 to 0.28 mm. Tabulae most often deeply infundibuliform, sometimes incomplete, forming 'dissepimental' structure near the corallite wall (Fig. 25G). In several places, concave tabulae are observed (see discussion below). In some corallites, thick- or thin-walled and shallow syrinx are clearly visible, centred or (more often) placed eccentrically (Fig. 25C–E). It occurs in most of corallites. Its diameter ranges from 0.24 to 0.42 mm. Tabulae rarely horizontal or meniscshaped. Septal spines are rare, conical, sharp and deeply embedded in stereoplasm. Tabular spines not observed. The wall microstructure is composed of concentrically arranged fibres, with traces of spines visible as bunches of radially arranged fibres (holacanths).

Measurements. Table 31.

Discussion. One of the specimens (ZPAL T.18-14/34) described above was originally assigned by Nowiński (1992) to *Armalites minimus* Nowiński, while the others to *Roemerolites lublinensis* Stasińska and Nowiński. All materials, the type specimen (the only specimen) of *A. minimus*, the holotype of *R. lublinensis* (ZPAL T.25 T.XI/129) and other material of '*R. lublinensis*', have been reinvestigated (three new thin sections were prepared of the holotype of *R. lublinensis* and four new of '*R. lublinensis*' from the Bolechowice Quarry). Observations from this material demonstrate that the specimens described by Nowiński (1992) as *R. lublinensis* and *A. minimus* are conspecific. They differ from *R. lublinensis* by the (1) presence of axial canal, (2) much thicker walls and (3) nearly absent cerioidal zones of corallum, which is common in *R. lublinensis*.

S. minimus shows very modest variation of corallite diameters (0.100 and 0.051 for two coralla). In contrast, the wall thickness is highly variable (variation coefficients 0.247 and 0.235, respectively), which however may be an effect of sectioning on different growth stages of the corallites, that is, in the proximal parts of corallum the walls are thick, and in the distal parts they are considerably thinner. Moreover, the variation coefficients for the measured characters are similar for two investigated coralla.

Occurrence. Central Kielce Subregion (middle/late Frasnian); Southern Kielce Subregion (late Frasnian).

Genus SYRINGOPORA Goldfuss, 1826

Type species. *Syringopora ramulosa* Goldfuss, 1826 from the Lower Carboniferous of Germany.

Diagnosis. See Tchudinova (1986, p. 80).

Remarks. This very common genus contains species of worldwide distribution, ranging stratigraphically from the Late Ordovician to Early Permian (Nowiński 1991).

FIG. 25. A–F, *Sapounofouskilites minimus* (Nowiński, 1992). Jaźwica/Góra Łgawa, Holy Cross Mountains, Poland; set R, Late Frasnian. Specimen ZPAL T.25 T. XVIII 14/34. A, transverse section. B, longitudinal section. C–F transverse sections. Note variable morphology and placement of axial canal (C–central, D, E–peripheral, F–absent) and highly variable wall thickness. G, *Sapounofouskilites minimus* (Nowiński, 1992). Panek Quarry, Bolechowice Holy Cross Mountains, Poland. Late Frasnian. Specimen ZPAL T.25 BP007. Longitudinal section. Arrows show dissepimental tabulae diagnostic for the genus. H–J, *Sapounofouskilites minimus* (Nowiński, 1992). Panek Quarry, Bolechowice Holy Cross Mountains, Poland. Late Frasnian Specimen ZPAL T.25 T. XVIII 2/3. H, transverse section. I, longitudinal section. J, longitudinal section, detail of I. Note dissepimental tabulae diagnostic for the genus. K, *Sapounofouskilites minimus* (Nowiński, 1992). Panek Quarry, Bolechowice Holy Cross Mountains, Poland. Late Frasnian. Specimen ZPAL T.25 T. XVIII 2/1. Longitudinal section. Note deep calyces with sharp edges. L–N, *Syringoporella raritabulata* Nowiński, 1992, holotype. Sowie Górki, Holy Cross Mountains, Poland; set G, Middle Frasnian. Specimen ZPAL T.25 T. XVIII 24/38. L, transverse section. M, longitudinal section. N, longitudinal section, detail of M showing up-arched connecting tubes. Scale bars represent 1 mm.

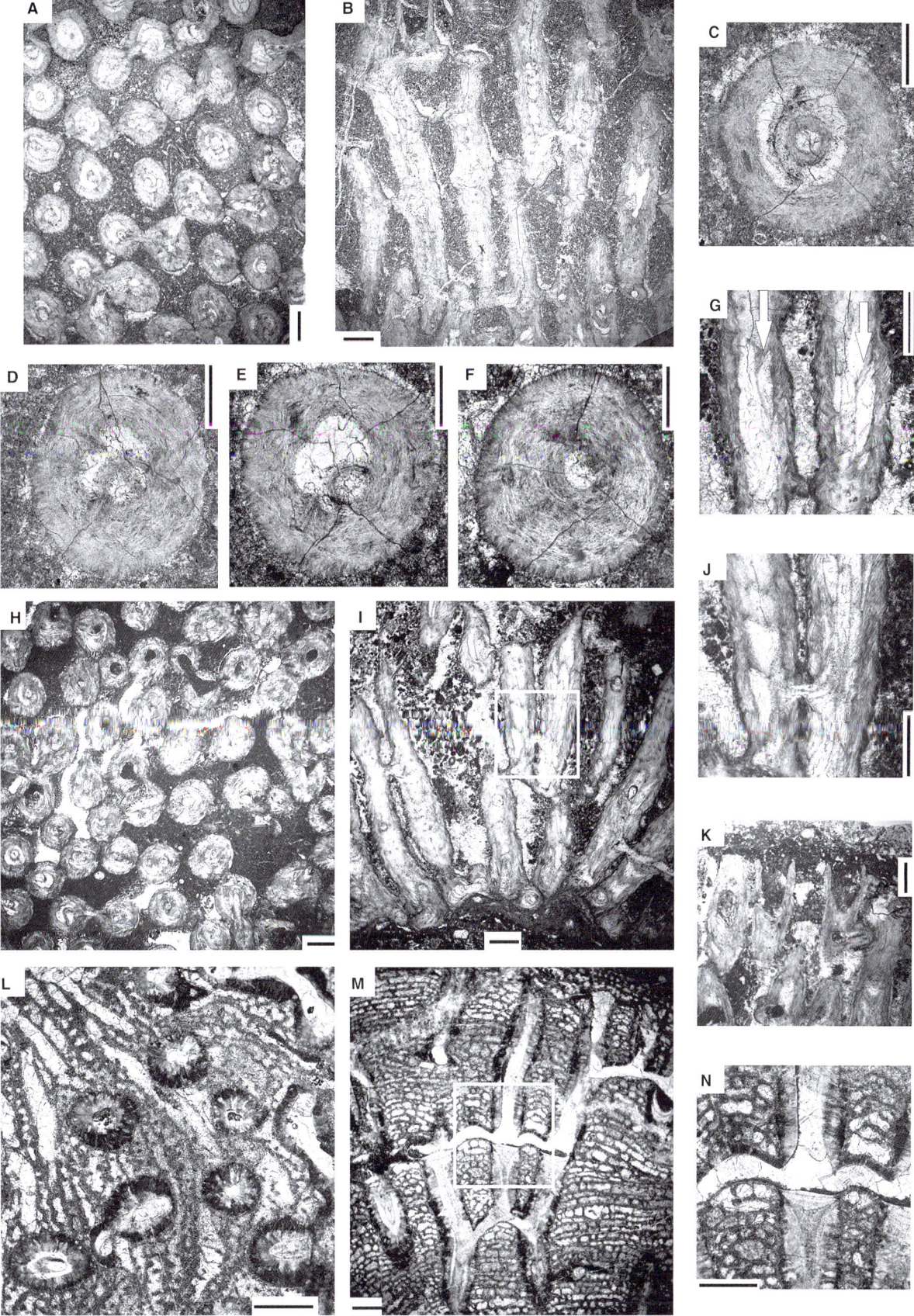

Syringopora cf. *volkensis* Tchernychev, 1938
Figure 24E–F

* ? 1938 *Syringopora volkensis* n. sp. Tchernychev,
            pp. 125–126, 140, pl. 5, fig. 2.
    1959 *Syringopora* cf. *volkensis* Tchernychev; Dubatolov,
            p. 194, pl. 61, figs 3, 4.
  ? 1987 *Syringopora* ex gr. *volkensis* Tchernychev; Hladil,
            p. 43.
  v. 1992 *Syringopora volkensis* Tchernychev; Nowiński,
            p. 199, fig. 13A–B.
  v. 2003 *Syringopora volkensis* Tchernychev; Nowiński,
            pp. 157–158, pl. 107, fig. 3.

Type material. Specimen 29/5256, coll. B. B. Tchernychev, Geological Museum VSEGEI, St. Petersburg, Russia (Tchernychev, p. 5, fig. 2).

Material. Jaźwica/Góra Łgawa (set R): Two coralla, four thin sections (ZPAL T.25 J024–026, J046).

Description. Coralla massive, up to 70 mm in diameter. Calyces deeper than wider, with infundibular bottom. Edges of calyces sharp. Corallites cylindrical, round or nearly round in cross section, 0.94–1.54 mm in diameter (mean 1.33 ± 0.16 mm, $N = 33$). They are closely spaced 0.02–0.18 mm, very often in contact. Walls uneven, 0.18–0.36 mm in thickness; epitheca not preserved. Tabulae strongly bent downwards, deeply infundibuliform, forming discontinuous axial canal, which is placed centrally or peripherally. The axial canal is sometimes continued through connecting tubuli to neighbouring corallites. Its diameter varies from 0.54 to 0.66 mm, and its internal diameter varies from 0.30 to 0.40 mm. In several cases, convex or incomplete tabulae are observed. Very short, wide at the base tabular spines occur randomly. Septal spines are few, randomly placed, 0.12–0.24 mm long and very sharp. Connecting tubuli few, 0.20–0.34 mm in diameter, their spacing unknown.

Remarks. The material described above shows a remarkable similarity with that described by Tchernychev (1938) from Vaygatch Island. The following major differences with the original material can be observed:
1. The corallite diameter in the type material is somewhat smaller, 0.9–1.2 mm. The range of diameter in the Holy Cross Material includes the values mentioned above; however, the mean diameter is somewhat larger.
2. The diameter of connecting tubuli, which is 0.5–0.8 mm in original material, is smaller, but this may be an effect of very few connecting tubuli being visible and only three measurements have been taken.
The discussed samples may be conspecific with the Frasnian material of *S. volkensis* described by Dubatolov (1959) from the Kuznetsk Basin. In the Dubatolov's material, however, the corallite diameters are slightly larger (1.1–1.3 mm), and the connecting tubes are smaller (0.5 mm) than in the type material.

Occurrence. Southern Kielce Subregion, Moravia (Frasnian), Kuznetsk Basin (Givetian). Hladil (1987) reported *S. ex gr. volkensis* from the Famennian of Moravia.

Syringopora? *tikhyiformis* (Stasińska and Nowiński, 1978)
Figure 24G–H

* v. 1978 *Aulocystis tikhyiformis* sp. n. Stasińska and
            Nowiński, p. 212, pl. 24, figs 1–3.
  v. 1992 *Aulocystis tikhyiformis* Stasińska and Nowiński;
            Nowiński, p. 200.
  v. 2003 *Aulocystis tikhyiformis* Stasińska and Nowiński;
            Nowiński, p. 164, pl. 113, figs 2, 3; pl. 114, fig. 1.

Holotype. Specimen ZPAL T.11/191.

Paratypes. Coll. Stasińska and Nowiński, Institute of Paleobiology, Warsaw (Stasińska and Nowiński, 1978 pl. 24, fig. 3).

Material. Jaźwica/Góra Łgawa (set R): One corallum, three thin sections (ZPAL T.25 J091–093).

Description. Corallum fascicular, moderately large, probably nodular, with the largest diameter (deduced from the thin section) about 50 mm. Corallites cylindrical, twisted, irregular. Corallite diameter variable, from 0.76 to 2.00 mm, most often 1.2–1.5 mm. Walls thick, even, single wall thickness 0.26–0.38 mm. Tabulae very thin, irregular, deeply concave, strongly inclined, very thin. Axial canal discontinuous. Connecting tubuli rare, thick (Fig. 24H). Septal spines sharp, short, small, dispersed. Tabular spines are rare, small, conical and sharp.

Remarks. Stasińska and Nowiński (1978) and later Nowiński (1992, 2003) assigned this species to the genus *Aulocystis*. However, it cannot be a member of this genus, of the family Aulocystidae, or even the order Auloporida, as there are – not numerous, but present – connecting tubuli. Stasińska and Nowiński (1978, p. 212) stated that tubuli are absent, but these can be observed on the holotype (Stasińska and Nowiński 1978, pl. 24, fig. 3) in the left part of the section. Therefore, on the basis of corallum structure, presence of connecting elements, shape of tabulae and presence of discontinuous axial canal, this species is assigned to the genus *Syringopora*.

Occurrence. Southern Kielce Subregion, Radom-Lublin area (Late Frasnian).

Family MULTITHECOPORIDAE Sokolov, 1950

Genus SYRINGOPORELLA Kettner, 1934

Type species. *Syringopora moravica* Roemer, 1883; (Givetian) Moravia.

Remarks. The genus consists of 10 species of worldwide distribution, ranging stratigraphically from Late Silurian to Early Carboniferous (Tchudinova 1986; Niko 2001). The *Syringoporella* species described below occur together with stromatoporoids, and this association is known as *Caunopora* Roemer (1844; the 'genus' *Caunopora* however was created by Phillips 1841). Most caunopores have been assigned to *Syringopora*, but Mistiaen (1984) proved that the microstructure is different from *Syringopora*. Association of *Syringoporella* species with stromatoporoids is quite common (Nowiński 1992; Niko 2001).

Syringoporella raritabulata Nowiński, 1992
Figure 25L–N

* v p 1992 *Syringoporella raritabulata* n. sp. Nowiński, pp. 200, 208, fig. 14A, B.
  v p 2003 *Syringoporella raritabulata* Nowiński; Nowiński, p. 159, pl. 109, fig. 1.

Material. Sowie Górki (set G): one specimen (holotype), two thin sections (ZPAL T.25 T.XVIII – 24/38).

Description. Small discoidal corallum, about 40 mm in diameter. The corallum is overgrown by a stromatoporoid *Stromatoporella* (det. B. Mistiaen). The corallites are short, cylindrical, weakly curved, spaced irregularly (0.14–2.00 mm). In transverse section, they are round, measuring 0.95–1.40 mm in diameter, rarely slightly oval. The corallite walls 0.20–0.50 mm thick, composed of two layers of fibro-radial microstructure. The internal layer appears to be lighter in polarized light, the external one darker; the border between them is unsharp. Epitheca thin, probably discontinuous, poorly preserved. Connecting tubes long, arched distally (Fig. 25M), with upper wall thicker (0.16–0.18 mm) and lower thinner (0.04–0.08 mm), with external diameter 0.28–0.40 mm. Tabulae very rare, most often funnel-shaped, forming sometimes centrally placed axial canal, or nearly vertical, forming a 'dissepimental zone' near the wall, rarely concave; commonly long segments of corallites are totally devoid of any tabulae. Septal apparatus poorly developed, usually as very rare, short, and very wide at the base, conical spines (buttons).

Discussion. Nowiński (1992) described this species based on two specimens. One specimen is excluded from *S. raritabulata* on the basis of having thicker walls and much better developed tabulae, and it is described below as *S.* sp. The description presented above has been supplemented by information on the (1) presence of axial canal, (2) presence of very rare septal spines and (3) differences in wall thickness in connecting elements.

S. raritabulata, similarly as species described below, overgrows a stromatoporoid. It seems that it had little negative influence on sponge growth, as the laminae near corallites are weakly bent downwards. As stated by Nowiński (1992), *S. raritabulata* differs from all other species of discussed genus by having much larger corallites.

Occurrence. Central Kielce Subregion (Middle Frasnian).

Syringoporella sp.
Figure 23F–G

* v p 1992 *Syringoporella raritabulata* n. sp. Nowiński, pp. 200, 208.
  v p 2003 *Syringoporella raritabulata* Nowiński; Nowiński, p. 159.

Material. Jaźwica (set K): one corallum, two thin sections (ZPAL T.25 J089–090); Wietrznia (Frasnian sets): Three thin sections (unknown number of corala; ZPAL T.25 STA W 011–013).

Description. Small discoidal corallum, about 40 mm in diameter. The corallum is overgrown by a stromatoporoid (Actinostromatidae gen. indet., det. B. Mistiaen). The corallites are short, cylindrical, weakly curved, spaced irregularly (0.80–2.20 mm). In transverse section, they are round, measuring 0.82–1.38 mm in diameter, rarely slightly oval. The corallite walls are 0.08–0.18 mm thick, composed probably of two layers of fibro-radial microstructure. Epitheca thin, probably discontinuous, poorly preserved. Connecting tubes rare, long, aberrant, with external diameter 0.40–0.90 mm. Tabulae numerous, deeply infundibuliform, forming centrally placed axial canal, or nearly vertical, forming a 'dissepimental zone' near the wall, rarely concave. Septal apparatus absent.

Remarks. The specimen from Jaźwica described here cannot be assigned to *S. raritabulata* (see Nowiński 1992, p. 208), because it has (1) much thinner walls and (2) much better developed tabulae. This species differs from all other known *Syringoporella* species by having much larger corallites (e.g. 0.4–0.7 mm in the type species, *S. moravica*; see Tchudinova 1986, p. 163). The specimen from Wietrznia is assigned to this species only tentatively, and it differs somewhat from the first specimen by having less densely spaced tabulae and less visible axial canal.

Occurrence. Southern Kielce Subregion (Frasnian).

Order AULOPORIDA Sokolov, 1952
(= AULOPORACEA Sokolov, 1952 *sensu* Hill, 1981)

Family AULOPORIDAE Milne-Edwards and Haime, 1851

Remarks. It should be added to the diagnosis for Auloporidae that members may form nonencrusting coralla (see Scrutton 1990).

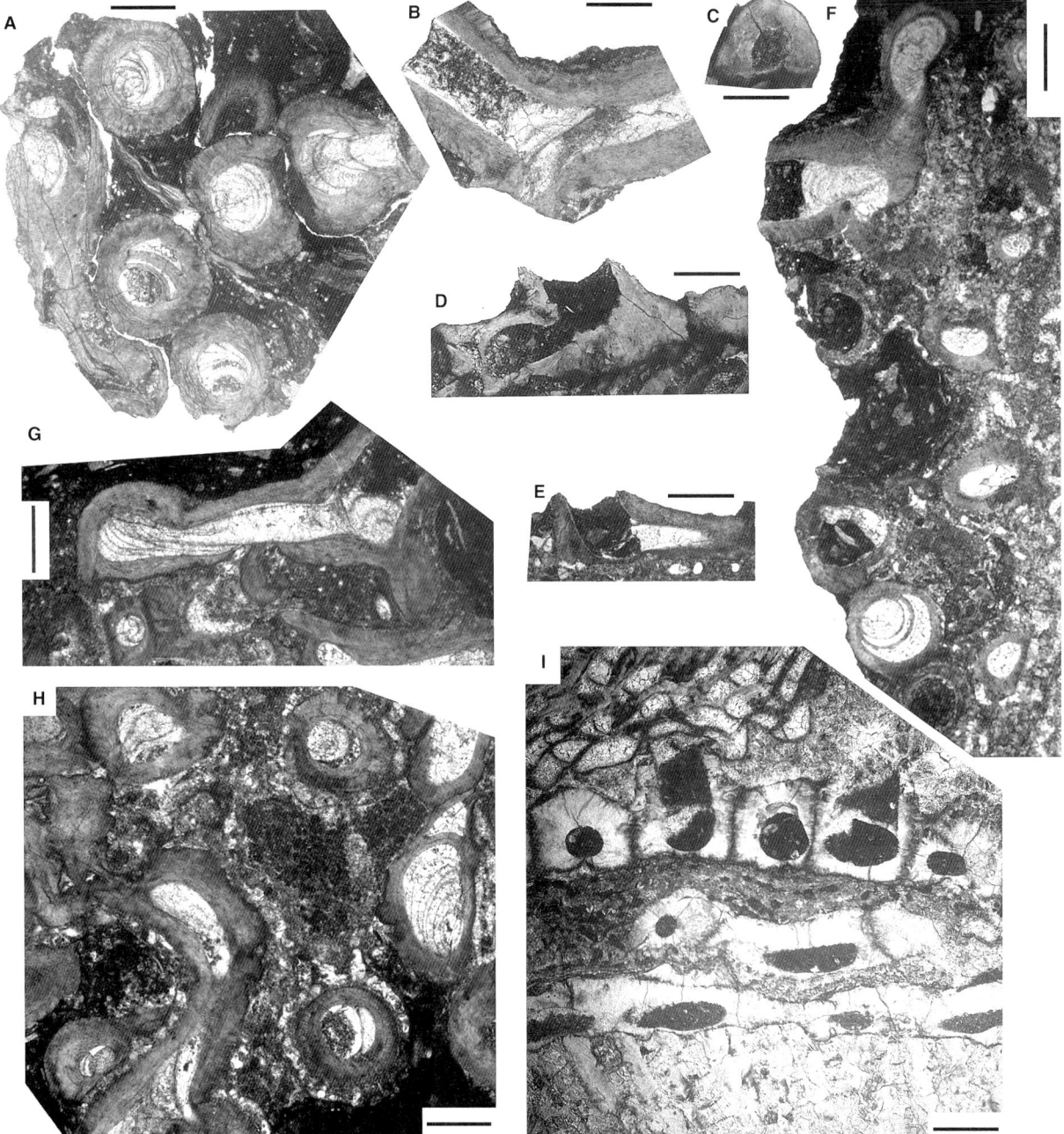

FIG. 26. A–B, *Aulopora* sp. B. Kowala Railroad Section, Holy Cross Mountains, Poland; set A, Early Frasnian. Specimen ZPAL T.25 K009–011. A, transverse section. B, longitudinal section. C–D, *Aulopora slosarskii* sp. nov. Kowala Railroad Section, Holy Cross Mountains, Poland; set A, Early Frasnian. Specimen ZPAL T.25 MZHT 210/D. C, transverse section. D, longitudinal section; see also Figure 23C. E–F, *Aulopora* sp. A Kowala Railroad Section, Holy Cross Mountains, Poland; set A, Early Frasnian. Specimen ZPAL T.25 KOW 11.002. E, longitudinal section. F, transverse section. G–H, *Aulopora* sp. C Kowala Railroad Section, Holy Cross Mountains, Poland; set A, Early Frasnian. Specimen ZPAL T.25 KOW 11.002. G, longitudinal section. H, transverse section. I, *Aulopora* cf. *compacta* Tchernychev, 1941. Trzemoszna, Holy Cross Mountains, Poland; Late Givetian. Specimen ZPAL T.25 TR 025. Scale bars represent 1 mm.

Genus AULOPORA Goldfuss, 1826

Type species. Aulopora serpens Goldfuss, 1826; Middle Devonian; Bensberg or Eifel, Germany.

Remarks. Aulopora is one of the most common tabulates in the Devonian; it is also one of the most often misidentified tabulate corals. The type species *A. serpens* is vaguely defined, but as the types of the subspecies are missing, a neotype should be established. Moreover, the type specimens of both subspecies of *A. serpens* may also be missing in the original Goldfuss collection at the university of Bonn (A. T. Halamski, pers. comm. 2003).

About 120 species have been established ranging from the ?Ordovician to Permian (Hill 1981) and with a world-wide distribution. However, most of them, except for the recently described ones, should be revised and rede-scribed, especially the species introduced at the end of nineteenth to the beginning of twentieth century are in need of revision.

Aulopora cf. *compacta* Tchernychev, 1941
Figure 26I

*? 1941   *Aulopora compacta* nov. sp. Tchernychev, p. 128, pl. 1, fig. 7, pl. 3.

? 1952   *Mastopora compacta* (Tchernychev); Sokolov, p. 156, pl. 40, figs 1–3.

? 1959   *Mastopora compacta* (Tchernychev); Dubatolov, p. 200, pl. 62, fig. 2.

? 1985   *Mastopora compacta* (Cernysev); Birenheide, p. 115, pl. 37, fig. 5, pl. 39, fig. 1.

? 1993a   *Aulopora* (*Mastopora*) *compacta* Tchernychev; May, p. 195, pl. 13, fig. 5, pl. 14, fig. 3.

? 1999   *Aulopora* (*Mastopora*) *compacta* Cernyšev; Brühl, p. 56, pl. 40, figs 188–191.

Type material. Specimen 121–11 (coll. Tchernychev), Palaeontological Institute RAS, Moscow, Russia (Tchernychev, pl. 1, fig. 7).

Material. Trzemoszna: One corallum, two transverse thin sections (ZPAL T.25 TR 024–025); Laskowa (set ?A): One corallum (ZPAL T.25 LAS R.049); Wietrznia: One corallum, three thin sections (ZPAL T.25 STA W 001–003).

Description. Coralla encrusting stromatoporoids and other tabulates, small. They are composed of closely adhering, compacted, closely adhering to each other corallites, most often rectangular in cross section. They are 1.20–1.60 mm wide and 1.10–1.40 mm high. Lumina round measuring 0.60–0.80 mm in diameter, or oval, measuring 0.34–0.70 × 0.54–1.20 mm. Calyces with rounded edges, about 1 mm in diameter, narrower than corallite. Walls very thick, with upper wall arched, and inner surface even. The wall of the upper side of corallite (thickest) varies between approximatively 0.40 and 0.60 mm, while the bottom wall is much thinner, 0.06–0.28 mm. Epitheca very thin, visible as thin dark median line between corallites, when they are in touch, often also on upper walls of individuals. Tabulae and septal spines are invisible on the material.

Discussion. The material described above resembles strongly the material of Tchernychev (1941), but without longitudinal sections, it is impossible to determine this

precisely. The material differs slightly from that described by Brühl (1999) by having larger corallite diameters and spines are not present; it differs from that described by May (1993a) by the absence of spines.

Some authors (e.g. Stasińska 1974; Hill 1981; May 1993a) consider *Mastopora* to be a distinct genus or a subgenus of *Aulopora*. Analysis of material from the Eifelian of the Łysogóry Region, as well as previously published (e.g. Mistiaen 1988), demonstrates that certain auloporids, especially of '*Mastopora*' type of organization, show extensive phenotypic plasticity. In large parts, they can be arranged in regular eyelets (lacunae), while at the peripheral zones they become much more compact, similarly, as recent scleractinians (Todd, 2008).

Occurrence. Kostomłoty Transitional Zone (Givetian), Northern Kielce Subregion (Frasnian), Main Devonian Field (Frasnian), Kuznetsk Basin (Frasnian), Rheinisches Schiefergebirge (Eifelian–Givetian).

Aulopora slosarskii sp. nov.
Figures 23C, 26C–D

Derivation of the name. In honour of Antoni Ślósarski (1843–1897), a Polish naturalist.

Type locality and horizon. Kowala Quarry, Holy Cross Mountains, Poland; lithological complex 'C', Kowala Fm., below *punctata* Zone, Frasnian.

Diagnosis. Coralla forming network. The lacunae (eyelets) elongated, measuring 4–6 × 7–9 mm. Corallites conical, 2.4–2.7 mm in length, up to 1.7 mm in diameter. Calyces round in cross section, with rounded edges, raised over substratum. Thickness of corallite walls variable, from 0.08 to 0.64 mm. Tabulae rare, dissepimenta-like.

Holotype. Specimen ZPAL T.25 MZHT 210/D (Figs 23C, 26C–D); corallum.

Paratype. Corallum from the type locality and strata, three thin sections (ZPAL T.25 KA 001–003).

Description. Coralla encrusting *Crassialveolites* and stromatoporoids. Coralla forming most often network, anastomosing, in some places corallites chaotically arranged. The maximal size of corallum is 60 × 50 mm. Lacunae slightly elongated, usually formed by 5–6 corallites, measuring 4–6 × 7–9 mm. Corallites slightly conical, reaching 4.5 mm in length, usually 2.4–3.0 mm; in cross section, round or slightly elliptical, flattened at the base. Their diameter in the place of budding varies from 0.9 to 1.1 mm, and the calyx diameter usually 1.4–1.7 mm. Calyces with rounded edges, raised over substratum. Walls with variable thickness, 0.08–0.64 mm, with the lower (adhering to the substratum) wall thinner. The internal surface of wall even.

Tabulae rare, oblique, adhering to the wall. Septal spines absent.

Remarks. A. *slosarskii* sp. nov. resembles, mostly by its biometrical characters, A. *lepida* Ermakova (Table 32). It differs from A. *lepida* by having shorter corallites and by the absence of the septal apparatus. The corallite length is similar to that of A. *venusta* Tchernychev, from which the new species differs by the diameters of calyces that are significantly larger. In *Aulopora heckeri* Tchernychev, the length of the corallites is similar, but the diameters of calyces are much smaller. It resembles also *Aulopora* sp. A described below but differs by having smaller eyelets of corallum and absence of the septal apparatus. *Aulopora* sp. C has a compact corallum of 'Mastopora' type, and in addition, the latter species has numerous tabulae.

The new species and already known *Aulopora* sp. from coeval strata are compared in Table 32.

Occurrence. Southern Kielce Subregion (Early Frasnian).

Aulopora sp. A
Figure 26E–F

Material. One corallum (six thin sections, ZPAL T.25 KOW 11.002.1–6) coming from the Kowala quarry, probably set C.

Description. Corallum encrusting *Alveolites*. Corallum forming regular network, with eyelets composed of 4–6 corallites. Corallites tubular or slightly conical in shape, 2.5–3.5 mm in length, round (0.70–1.50 mm in diameter) or oval (0.80–1.00 × 0.70–1.10 mm) in cross section, often flattened at the base. Calyces raised over substratum, with slight thickening of the walls. Walls even, highly variable in thickness extending from 0.10 to 0.80 mm, but most often from 0.14 to 0.22 mm. Tabulae rare, strongly inclined, funnel shaped or convex, in one place forming axial canal. Septal spines very rare, observed only on one section in new offset.

Remarks. *Aulopora* sp. A is similar to A. *gratus* Ermakova from the Frasnian of the Central Devonian Field in Russia; it differs by having slightly shorter corallites and a different corallum structure.

Occurrence. Southern Kielce Subregion (Frasnian).

Aulopora sp. B
Figure 26A–B

Description. Corallum very small, branching. Corallites round, very large, 1.96–2.20 mm in diameter. Corallite walls are 0.40–0.60 mm thick (SWT), even and composed of two layers: inner one concentric-lamellar, the outer one radio-lamellar, each of them consisting about 50 per cent of the wall thickness. Epitheca invisible. Tabulae inclined or infundibuliform of very variable thickness, sometimes forming axial canal. Septal apparatus absent.

Remarks. The corallum described above resembles the corallum of A. *levatus* Ermakova (Ermakova 1958, p. 34, pl. 6, figs 1–5) from the Frasnian of Voronezh Region. It is similar by calyce (corallite) diameter but differs from the latter species by having much thicker walls (0.2–0.3 in A. *levatus*).

TABLE 32. Comparison of Frasnian species of *Aulopora.*

Species	Corallite length	Calyx diameter	Wall thickness	Bibliography
A. *cylindrica* Tchernychev	0.75–1.50	0.50–0.65	?	Sokolov (1952)
A. *compacta* Tchernychev	2–4	1–2	?	Tchernychev (1941)
A. *flexuosa* Ermakova	Curved	2.0–2.5	?	Ermakova (1958)
A. *gratus* Ermakova	3.5–6.0	0.70–0.80	?	Ermakova (1958)
A. *heckeri* Tchernychev	1.5–3.0	0.4–0.6	?	Sokolov (1952)
A. *lepida* Ermakova	1.5–2.0	1.3–1.6	0.2	Ermakova (1958)
A. *levatus* Ermakova	5.0–7.0	2.0–3.0	0.2–0.3	Ermakova (1958)
A. *minuicompacta* (Ermakova)	1.0–3.0	0.70–1.35	0.30–0.40	Ermakova (1958)
A. *nana* (Ermakova)	1.0	0.6–1.0	?	Ermakova (1958)
A. *parvulus* (Ermakova)	?	0.80–1.20	?	Ermakova (1958)
A. *schelonica* Tchernychev	7.0–8.0	2.0	?	Sokolov (1952)
A. *slosarskii* sp. nov.	2.4–3.0	1.4–1.7	0.08–0.64	This paper
A. *venusta* Tchernychev	2.0–3.5	0.8–1.0	?	Sokolov (1952)
A. *verticillata* Sokolov	3.0–5.0	1.0–1.2	?	Sokolov (1952)
A. sp. A	2.5–3.5	0.7–1.5	0.10–0.80	This paper
A. sp. B	?	1.96–2.20	0.40–0.60	This paper
A. sp. C	?	1.30–1.60	0.14–0.50	This paper

All values in millimetres.

Material. One small, incomplete, poorly preserved corallum from the biostromal complex (set A) of Kowala Quarry, Frasnian; four thin sections (ZPAL T.25 K009–011).

Occurrence. Southern Kielce Subregion (Frasnian).

Aulopora sp. C
Figure 26G–H

v ? *1992 Aulopora* sp. (cf. *A. verticillata* Sokolov); Nowiński, p. 200.

Material. Kowala Quarry: One complete, well-preserved corallum from the rubble near biostromal complex (set A) five thin sections (KOW R.072B.1–5).

Description. Corallum very small, encrusting. It is composed of chaotically arranged corallites, often closely adhering to each other (as in '*Mastopora*'). Corallites round, oval and subtriangular in cross section, 1.30–1.60 mm in diameter. Corallite walls very variable, 0.14–0.50 mm (SWT) in thickness, thicker in proximal parts of corallites, distally tapering, even, composed of two layers: inner one concentric-lamellar, the outer one radio-lamellar, each of them consisting about 50 per cent of the wall thickness. Epitheca very thin. Tabulae numerous, strongly inclined, rarely infundibuliform, usually very thin (in few places slightly thicker tabulae may be observed). They rarely form an axial canal. Septal apparatus absent.

Remarks. This specimen resembles *Aulopora lepida* Ermakova from the Frasnian of the Voronezh Region, but it differs by its more chaotic arrangement of corallites and the absence of septal spines. Thin sections referred by Nowiński (1992) to *Aulopora* sp. (cf. *A. verticillata* Sokolov) may belong to this species, but the specimen is neither cut perpendicularly or longitudinally.

Occurrence. Southern Kielce Subregion (Frasnian).

Family AULOCYSTIDAE Sokolov, 1950

Genus ADETOPORA Sokolov, 1955

Type species. A. humilis Sokolov, from the Upper Carboniferous of western slopes of Ural Mountains.

Remarks. This little known genus ranges stratigraphically from Silurian (Ospanova and Leleshus 1988) up to Upper Carboniferous (Sokolov 1955). Probably up to seven species are known. Hill (1981, p. F642) had doubts concerning the validity of the genus and indicated the possible homonymy with *Aulocystella* Kuzina *in* Sokolov 1955. *Adetopora* differs, aside from other characters, by the absence and/or poor development of spines.

Adetopora? *tikhyi* (Sokolov, 1952)
Figure 23D–E

* 1952 *Aulocystis tikhyi* n. sp. Sokolov, p. 157, pl. 40, figs 4–6.
 1983 *Aulocystis tikhyi* Sokolov; Kulicka and Nowiński, p. 486, pl. 41, fig. 1.
 v. 2003 *Aulocystis tikhyi* Sokolov; Nowiński, p. 164, pl. 113, fig. 1.

Type material. Specimen 1813-1, All-Russia Petroleum Research Exploration Institute, St. Petersburg, Russia (Sokolov, 1952, pl. 40, figs 4–5).

Material. Kowala Quarry (sets A–C): four coralla, 12 thin sections (ZPAL T.25 KQ001–012); Jaźwica Quarry (set I): one corallum, four thin sections (ZPAL T.25 J 097–100).

Description. Coralla bushy, small, probably up to about 50 mm of the largest diameter. Corallites cylindrical, irregularly bent, rarely straight, round or nearly round in cross section, measuring 0.92–2.0 mm in diameter, most often 1.2–1.5 mm. They are spaced chaotically, sometimes wide apart, sometimes in contact. Calyces, deep, with sharp edges, walls near the edges thin. Walls very thick, 0.18–0.56 mm in thickness, most often 0.30–0.40 mm, with internal surface smooth and even. Wall microstructure concentric, lamellar. In certain corallites, the innermost layer of the wall seems to have radial microstructure. Epitheca irregularly wrinkled, thin. Tabulae very thin, slightly convex, sometimes incomplete, deeply infundibuliform, often parallel to the corallite wall. They form usually eccentrically, rarely subaxially placed, discontinuous syrinx. Septal and tabular spines absent, connecting elements not observed.

Discussion. Specimens belonging to this species have been assigned to the genus *Aulocystis* since the original description of Sokolov (1952). All authors describing it (e.g. Kulicka and Nowiński 1983; Nowiński 2003) were emphasizing the absence of septal and tabular spines. Also, the original description of Sokolov (1952, p. 157) contains statement about the absence of spines. However, all authors giving diagnosis to the genus *Aulocystis* (e.g. Sokolov 1962*a*, *b*; Hill 1981; Byra 1983; Birenheide 1985) were evoking presence of septal spines as an important generic feature. Therefore, this species cannot be assigned to the genus *Aulocystis*.

The presence of following features: irregularly spaced corallites, absence of connecting elements and concave or infundibuliform tabulae permits to assign the investigated specimens to the family Aulocystidae. The only member of this family that does not possess any spines is genus *Adetopora* Sokolov. The discussed species is tentatively assigned to this genus. For definitive generic determination, the type material of Sokolov must be investigated.

Occurrence. Southern Kielce Subregion (Early Frasnian), Main Devonian Field, Western Periuralia and Tatarstan (Frasnian).

INTRACOLONIAL VARIATION

Intraspecific and intracolonial variation in anthozoans is well recognized in rugose and scleractinian corals (e.g. Best *et al.* 1984; Foster 1984; Fedorowski 1989; Höfling 1989; Oliver 1989; Sorauf and Mackey 1989; Sutherland 1989), while in the tabulate corals only favositids and heliolitids have been subject of systematic studies. Researchers of the tabulate corals dealt with general aspects of variability, that is, describing qualitatively patterns of variation (e.g. Pandolfi 1989; Scrutton 1989), whereas quantitative analysis of certain groups has not been executed. However, several studies on intracolonial and intraspecific variability, principally in two above-mentioned groups are published by Young and Noble (1989), Hubmann (1991*b*, 1992), Young and Elias (1997), Fernández-Martínez and Mistiaen (2003), Plusquellec *et al.* (2004), Tsukada (2005) and Mõtus (2006).

Analysis of variation is very important in taxonomical studies and can be effectuated in one of the following ways:

1. In order to recognize variation of certain parameters within a given taxon.
2. In order to recognize variation of a given parameter in various taxa with the purpose to evaluate the value of the parameter as a criterion to distinguish separate taxa.

The major part of existing studies assumed that a character is taxonomically diagnostic and then investigated its variation in order to describe a given taxon (e.g. Iven 1980; Noble and Young 1984). Studies dealing with the characters themselves and investigating their taxonomic usefulness independently of systematic descriptions of tabulate corals are very few (e.g. Sutton 1966; Scrutton 1981; Lee and Noble 1988).

The value of selected biometrical parameters as taxonomic discriminators is investigated here, while intracolonial variation in order to describe taxa has been treated in the Systematic Part. Information on quantitative intracolonial variability is given in the tables included in the specific descriptions (see Systematic Palaeontology) or directly in the text. Analysis of several parameters commonly used for distinguishing between taxa is presented below. Several biometrical parameters are analysed here in order to evaluate their usefulness in taxonomic studies; these are lumen diameter (maximal and minimal), double wall thickness, pore diameter and tabulae spacing. Pore spacing was not analysed, as the data are very fragmentary and do not permit to draw broader conclusions. The groups under study are the Alveolitidae and the Coenitidae, with a special emphasis

on the former family (in addition, a single species of Syringoporidae is analysed).

Results

The coefficients of variation (V) of all investigated specimens are given in Table 33. In general, the V values for the different parameters are ranging from 0.051 to 0.609. The range and mean V values in alveolitids are as shown in Table 34.

Discussion

The analysed parameters in the investigated coralla have different variation ranges. The maximal and minimal lumen width is the parameter with lowest variation in both Alveolitidae (mean V values 0.160 and 0.155, respectively) and Coenitidae (Max LD = 0.124–0.229, Min LD = 0.154–0.189). However, a syringoporid, *S. minimus*, shows much smaller intracolonial variation ($V = 0.051$–0.100). These values are very similar to those known from Silurian heliolitids (Mõtus 2006), where the tabularium diameter (parameter analogous to the lumen diameter) is similarly variable as in *S. minimus* (for heliolitids: $V < 0.09$). It may be noted that Silurian *Avicenia* has higher intracolonial variation values (tabularium: $V = 0.1$ to 0.21; pore diameter $V > 0.34$; Zapalski and Nowiński 2011). Thus, alveolitids have higher intracolonial variation than heliolitids. Coenitids have slightly higher variation of the lumen diameters; their mean V value is close to 0.2.

The variation (V values) of pore and corallite diameters in the Alveolitidae is similar for both parameters and ranges from 0.101 to 0.194. Double wall thickness and tabulae spacing are the most variable characters in all investigated coralla. In alveolitids, the mean V value is 0.277. The wall thickness is less variable in Silurian heliolitids, where the highest observed V value was 0.247 for *Heliolites interstinctus* L. (Mõtus 2006). The double wall thickness and tabulae spacing in coenitids show similarly variation as in the alveolitids.

The parameters with the lowest intracolonial variation (low V value), such as corallite (lumen) diameter or pore diameters, seem to be good species discriminators. Such parameters are relatively constant, with majority of values occurring close to the mean value. Also, previous observations of Sutton (1966) and Powell and Scrutton (1978) show that lumen diameters (or corallite diameters) and pore diameters are good distinguishing features (see also Zapalski 2007*c*). Low variation of V values can be interpreted as a low susceptibility for environmental changes. However, Lee and Noble (1988) on the basis of favositid material concluded that the corallite size (corallite area

TABLE 33. *V* values of biometrical parameters in selected tabulate taxa.

Species	Sample/ThSe No.	Max LD	Min LD	DWT	PD	PS	TS
Alveolitidae							
A. suborbicularis	KOW R.060	0.174	0.168	0.376	0.126	–	0.325
A. suborbicularis	KOW 4.001	0.093	0.122	0.206	0.129	0.261	0.265
A. suborbicularis	KOW 9.A01	0.128	0.121	0.182	0.101	–	0.344
A. suborbicularis	J009–011	0.174	0.173	0.276	0.174	–	0.360
A. suborbicularis	JAZ 006	0.114	0.147	0.267	–	–	0.170
A. suborbicularis	KOW R.061	0.151	0.137	0.256	–	–	0.286
A. suborbicularis	GC007–010	0.195	0.176	0.286	–	–	0.331
A. compressus	KOW AX.001	0.131	0.113	0.340	0.149	0.197	0.292
A. maillieuxi	MRHN 7069	0.144	0.197	0.331*	–	–	0.207
A. maillieuxi	KOW 9.003	0.160	0.161	0.316	0.121	0.192	0.289
A. maillieuxi	J065–067	0.196	0.191	0.243	–	–	0.289
A. multispinosus	KOW 4.004	0.196	0.191	0.243	–	–	0.289
A. parvus	MRHN 2	0.145	0.131	0.207	–	–	–
A. parvus	MRHN 624	0.126	0.113	0.246	–	–	0.286
A. parvus	SG100–103	0.167	0.159	0.236	–	–	0.446
A. parvus	L015–017	0.159	0.104	0.196	–	–	0.279
A.? obtortiformis	MRHN 1806	0.104	0.137	0.247	–	–	0.609
A.? obtortiformis	MD 005–007	0.124	0.120	0.234	–	–	0.211
A.? tenuissimus	MRHN 1212	0.126	0.178	0.329	–	–	–
A.? tenuissimus	J006–007	0.105	0.148	0.532	–	–	–
A.? sp.	LAS 100	0.153	0.168	0.356	–	–	0.387
C.? multiperforatus	MRHN 4100a	0.153	0.149	0.318	0.146	0.199	0.232
C.? multiperforatus	PW020–024	0.127	0.135	0.305	0.163	–	0.278
C.? multiperforatus	ST 001–003	0.188	0.163	0.288	0.143	–	0.386
C. crassus	MRHN 90a	0.214	0.248	0.297	0.193	0.313	0.259
C. crassus	C010–011	0.160	0.147	0.205	0.185	–	0.225
C. aff. crassus	PW010–016	0.161	0.164	0.304	–	–	0.399
C. cavernosus	MRHN 1155	0.200	0.191	0.280	–	–	0.320
C. cavernosus	PW029–030	0.225	0.139	0.281	–	–	–
C. oliveri	MZHT210/D	0.171	0.134	0.213	0.194	0.142	0.347
A. fecunda	MRHN a417	0.189	–	0.330	0.180	0.345	0.333*
A. fecunda	TR002–003	0.173	–	0.156	–	–	0.291
A. fecunda	PL009–011	0.207	–	0.247	0.162	–	0.372*
A. fecunda	M001–003	0.192	0.181	–	–	–	0.247
Coenitidae							
P. escharoides	LAS 006	0.124	0.175	0.117	–	–	–
R. heuvelmansi	LAS R.001	0.216	0.154	0.284	–	–	–
R. heuvelmansi	L091–092	0.218	0.159	0.272	–	–	0.364
R. sp. A	LAS R.044	0.229	0.189	0.356	–	–	0.279
R. sp. B	LAS R.025	0.184	0.180	0.349	–	–	–
Syringoporidae							
S. minimus	J096	0.100	–	0.247	–	–	–
S. minimus	TXVIII–2/3	0.051	–	0.235	–	–	–

Measurements in millimetres. For abbreviations, see Table 1. ThSe, thin section. *One of two values, for example, tabulae spacing in axial-peripheral zones; the lower one was used.

measured in cross section) is not good species discriminator owing to its high variability. Nevertheless, the taxonomic value of a given parameter can be different in different groups of corals – the bimetrism in favositids, causing high variation, is a well-known feature (Scrutton and Powell 1980), while alveolitids do not seem to be the case. Conversely, the parameters with high *V* values, such as double wall thickness and tabulae spacing, are bad taxonomical discriminators in comparison with those discussed above.

The *V* values for alveolitids can be arranged from the lowest to the highest (all values for all parameters are treated together). The analysis of percentiles of *V* values in alveolitids (Fig. 27) shows that all *V* values can be

TABLE 34. Maximal, minimal and mean *V* values for selected parameters (Alveolitidae).

Parameter	Minimal *V* value	Maximal *V* value	Mean *V* value
Max LD	0.093	0.225	0.160
Min LD	0.104	0.248	0.155
DWT	0.156	0.532	0.277
PD	0.101	0.194	0.155
PS	0.142	0.345	–
TS	0.170	0.609	0.308

divided into three groups. About 55 per cent of them do not exceed 0.200, another 42 per cent of the *V* values are ranging between 0.200 and 0.399, and the remaining 3 per cent are above 0.399. Therefore, the group of lowest values can be categorized as normal intracolonial variation, the second group as moderate intracolonial variation and the last one as high intracolonial variation. Such a division with border values as stated above may be applied only to alveolitids; it is likely that for other groups of tabulates, the distribution of *V* values will be different as may be seen on the example of syringoporid discussed here.

In general terms, alveolitids and coenitids show higher intracolonial variation than syringoporid analysed here or heliolitids. Higher variation of two analysed groups can be caused by specifics of given group, as well as by environmental conditions. The different *V* values of specimens coming from the similar facies, but belonging to different groups, for example *S. minimus* and *A. maillieuxi* from Jaźwica, can suggest faint environmental influence on variation, with stronger genetical control on this phenomenon.

Also closely related species, for example, *A. multispinosus*, *A. parvus* and *A.* sp. coming from different facies often have similar *V* values, which can suggest genetical rather than environmental control on the variation, because they were growing under different environmental conditions. Similar values obtained for Belgian and Polish specimens within each given species may be caused either by genetic or environmental controls, as they come from similar facies. On the other hand, Watkins (2000), who on the basis of study of several groups of tabulates, that is, favositids, alveolitids, halysitids, heliolitids and syringoporids, showed that the variation of corallite diameter and spacing is related to the environment, and also Mõtus (2001, 2006) considered intracolonial variations as environmentally controlled.

Favositids (Favositina) are not useful as basic reference group for such studies. They show strong bimetrism, and apparently, their pattern of variation is not the same. Furthermore, their pattern of growth is different than that of alveolitids. Also various groups of corals react differently to the environment.

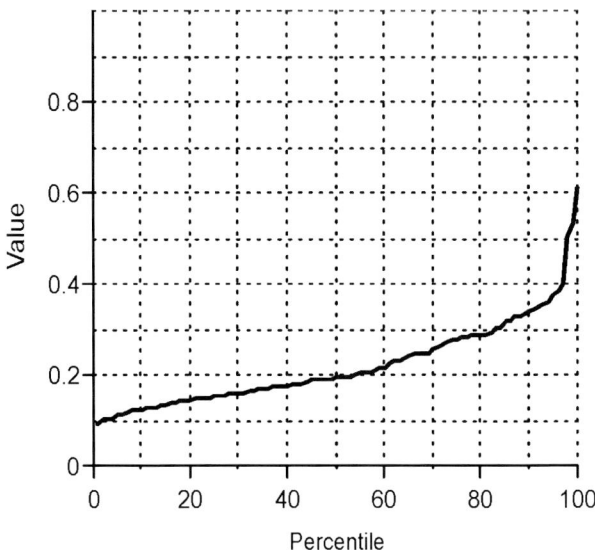

FIG. 27. Distribution of *V* values in Alveolitidae; *V* values for all parameters are treated collectively.

PARASITISM IN TABULATE CORALS

Tabulate corals are often associated with other organisms (e.g. Klaamann 1959; Schindewolf 1959; Hubmann 1991*a*: Tapanila 2004; Zapalski *et al.* 2008), and such an association is known since the description of *Pleurodictyum* occurring together with a 'worm' *Hicetes* (Goldfuss 1829). Also other organisms placed within the coralla are known since the papers of Sokolov (1948, 1962*b*). They were for a long time considered as 'commensals' (e.g. Plusquellec 1968; Oekentorp 1969; Tapanila 2005). Commensalism, however, occurs very rarely in modern interactions between organisms and cannot be recognized in ancient communities (Zapalski 2011). This interaction requires positive effect on one organism and neutral for the other, and absence of an interaction cannot be proven (Zapalski 2011).

First published mention of parasitism on tabulate corals was that by Stel (1976, 1978), but without any proof; similarly, Bertrand *et al.* (1993) used the term 'parasitism' without any discussion or justification. Organisms commonly infesting Palaeozoic corals, such as *Torquaysalpinx*, were also found in stromatoporoids. These were also traditionally interpreted as commensals (Tapanila 2005), but recent analyses of growth inhibition show that they were parasites of stromatoporoids (Zapalski and Hubert 2011).

The parasitic relationship between the tabulate coral *Favosites goldfussi* d'Orbigny and infesting organism *Chaetosalpinx ferganensis* Sokolov was analysed by Zapalski (2004, 2007*a*, 2009). The evidence for such an interaction is based on the following:

1. Presence of *Chaetosalpinx* between corallite walls or sometimes within septa, which leads to interpretation that these organisms must have perforated soft tissue of coral.
2. Host's phenotype modification, visible as folding and thickening of the corallite wall.
3. Long-term relationship between an infesting organism and coral: according to Tapanila (2002), at least 4 years. Combes (2001) indicates long-term interaction as an important indication of parasitism.
4. Possible inhibition of the host's growth.

It can be added here and following Littlewood and Donovan (2003) that these organisms meet the criteria of the definition 'host as habitat', which also indicates parasitism.

The analysis of Zapalski (2007*a*) was based on specimens from the Emsian–Eifelian of Grzegorzowice (Łysogóry Region) with *Chaetosalpinx ferganensis*, which infested *Favosites goldfussi*. The possibility of parasitic relationship between other tabulates (hosts) and associated taxa such as *Helicosalpinx* Oekentorp or *Actinosalpinx* Sokolov was indicated by Zapalski (2007*a*).

Chaetosalpinx? *plusquelleci* isp. nov., co-occurring with *Pachyfavosites polonicus* and *Alveolites multispinosus* are described and analysed, as well as *Helicosalpinx* cf. *asturiana* and *Helicosalpinx* isp. infesting *Alveolitella fecunda*. This is also the first report of *Helicosalpinx* from Poland.

Material

The material analysed here consists of two coralla of *Pachyfavosites polonicus* Nowiński from the Frasnian of Sowie Górki (six thin sections), two coralla of *Alveolitella fecunda* (Lecompte) from the Givetian of Trzemoszna (three thin sections) and one corallum of *Alveolites multispinosus* Dubatolov.

SYSTEMATIC ICHNOLOGY

At the beginning, the described organisms were classified as 'worms' (e. g. Plusquellec 1968) or as serpulids (Howell 1962). Tapanila (2002) classified these structures within ichnotaxonomy as new ethological category, *impedichnia*. Such a point of view was criticized by Bertling *et al.* (2006), who rejected these fossils from ichnotaxonomy. However, these structures are better considered ichnofossils, as they are not anatomical structures themselves, but only imprints on the tissues of other organisms. Therefore, the classification of Tapanila (2002) is followed here.

Ichnogenus CHAETOSALPINX Sokolov, 1948

Type species. *Chaetosalpinx ferganensis* Sokolov, 1948 from the Silurian of Fergana (Uzbekistan).

Diagnosis. See Tapanila (2002).

Remarks. The traces of *Chaetosalpinx* endobionts appeared in the Ashgillian, and the youngest known until now are Famennian. Hosts of species belonging to this ichnogenus include stromatoporoids, rugose and tabulate corals (Tapanila 2005; Zapalski 2007*a*; Zapalski *et al.* 2008).

Chaetosalpinx? *plusquelleci* ichnosp. nov.
Figure 28C–F

v. *1992* unnamed 'commensal' tubes Nowiński, p. 202, text-fig. 2A, B.

Derivation of the name. In honour of Dr. Yves Plusquellec, an outstanding tabulate researcher.

Type locality and strata. Sowie Górki, western quarry, Holy Cross Mountains, Poland. Upper Sitkówka Beds, set G of Racki 1992*a*, Middle Frasnian.

Type material (Fig. 28C–F). It was necessary to establish syntypes because it is not possible to have longitudinal and transverse sections of the same individual (Fig. 28C–F), which are the organisms infesting *Pachyfavosites polonicus* Nowiński (holotype of this species), specimen ZPAL T.25 T. XVIII 24/2 (four thin sections).

Other material. Sowie Górki: one corallum of *Pachyfavosites polonicus* Nowiński (two thin sections) coming from the Upper Sitkówka Beds (Middle Frasnian), infested by *Chaetosalpinx*? *plusquelleci* isp. n., specimen ZPAL T.25 SG 151–152; Kowala Railroad Section: one tube in the corallum of *Alveolites multispinosus* Dubatolov (Early Frasnian).

Description. Tubes of *Chaetosalpinx*? *plusquelleci* isp. nov. occur irregularly throughout the host corallum. The tubes occur usually between the corallite walls, but they often occupy the lumen of corallites (Fig. 28D). When placed between the corallite walls, they do not have their own walls. When occurring within the corallite lumina, a very thin wall occurs, with the microstructure resembling this of the host coral (see discussion below). They are most often round in cross section, very rarely slightly oval, with internal surface smooth and even. Their internal diameter ranges from 0.06 to 0.52 mm (mean 0.343 ± 0.1327 mm, $N = 38$), but a weak bimodal distribution of diameter size can be seen (Fig. 29). No internal structures have been observed. A strong anatomical influence on the host coral is visible on a cross section, where the anatomical structures of the host (tabulae) are attached to the 'worm' (Fig. 28D, arrow).

Remarks. The specimens of the new species have been tentatively assigned to genus *Chaetosalpinx* on the basis of lack of internal structures and absence of own wall.

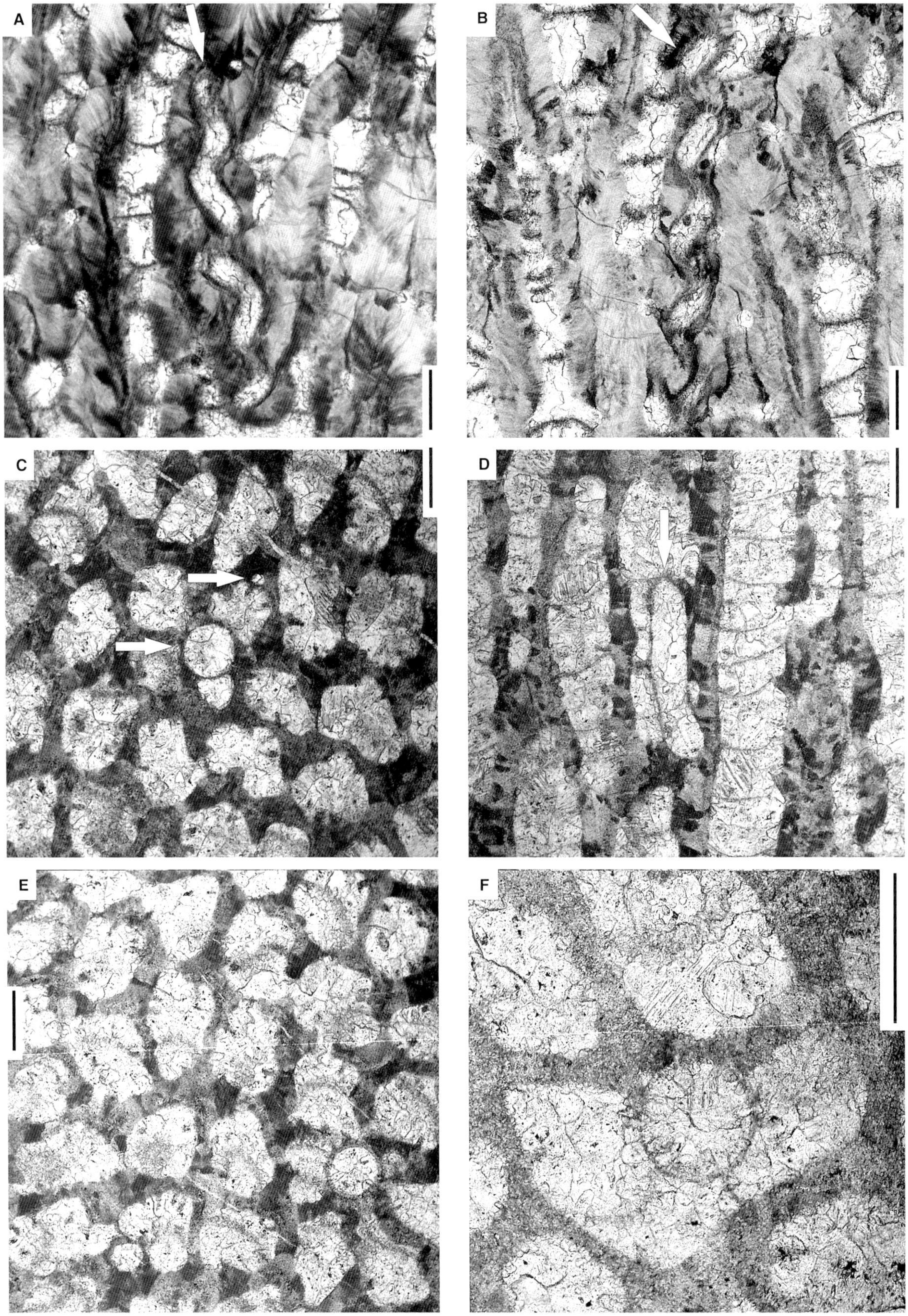
A
B
C
D
E
F

The individuals of the new species sometimes seem to have a wall, but its microstructure (observed in normal thin sections, under polarized light) seems to be the same as the host's wall. Thus, it is rather an 'encystation' of the infesting organism by the host coral than its own wall. Also, Plusquellec (1968), on the basis of type material of *C. australiensis* (Sokolov), concluded that the structure that seemed to be own wall of infesting organism (structural fossil) was in fact secreted by the host (bioimmuration). From the typical representatives of the genus *Chaetosalpinx*, that is, *C. ferganensis*, the described material differs in occurrence, sometimes within the corallite lumina.

The bimodal distribution of tube lumina diameters may be interpreted as change of tube diameter, similarly as in *Chaetosalpinx rex* Tapanila. The thin tube occurring between walls suddenly becomes larger – the intermediate sizes are rare, because this part of tube is much shorter than other parts of the tube. The small number of the smallest diameters may be an effect of location of the thin sections – close to the corallum surface. The new species is most similar to *Ch. australiensis* (Sokolov), from which it differs by the absence of horizontal elements within the tube (in fact, presence of this feature shows that *Ch. australiensis* is probably not *Chaetosalpinx*). From other known representatives of this genus, the new species differs by larger diameters and placement also within the corallite lumina. The Ordovician *C. rex* has lens-shaped tubes (in cross section) near the surface of host corallum, while the new species is uniformly round. From *C. sibiriensis* Sokolov, it differs by the absence of tabulae.

Occurrence. Central and Southern Kielce Subregion (Givetian–Frasnian).

Ichnogenus HELICOSALPINX Oekentorp, 1969

Type species. Helicosalpinx concoenatus (Clarke, 1908) from the Late Silurian of New York, USA.

Diagnosis. See Oekentorp (1969).

Remarks. Up to now, only two species have been assigned to the genus *Helicosalpinx* – the type species and the

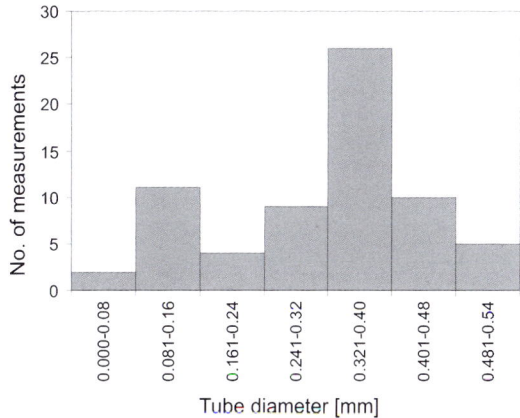

FIG. 29. Distribution of internal tube diameters of *Chaetosalpinx? plusquelleci* isp. nov. occurring in two coralla of *P. polonicus* Nowiński, coming from Frasnian of Sowie Górki (see description for further details) (N = 67).

Devonian *H. asturiana* Oekentorp. The discussed genus has been reported from the hosts belonging to different tabulate corals (alveolitids, favositids, theciids, calapoecids) as well as rugosans and stromatoporoids, coming from various strata (Ordovician–Devonian) with worldwide distribution (Tapanila 2005, table 1). This is the first report of *Alveolitella* hosting *Helicosalpinx*, as well as the first report of *Helicosalpinx* itself from Poland.

Helicosalpinx cf. *asturiana* Oekentorp, 1969
Figure 28A

Material. A corallum of *Alveolitella fecunda* (one thin section, ZPAL T.25 TR034) infested by four individuals, coming from the Givetian of Trzemoszna.

Description. Helicoidally curved tubes placed between the corallite walls. Within the 3- to 5-mm-long longitudinal section, five whorls are visible. The coiling is irregular and loose, with the radius of coiling probably smaller than the diameter of tube. Their maximal diameter is about 0.20 mm. Cross section unknown. The tube does not have its own wall.

Remarks. The described specimens are most similar to *H. asturiana*, but small material does not allow to measure certain parameters (such as angle of coiling) and to determine it precisely.

FIG. 28. A, *Helicosalpinx* cf. *asturiana* Oekentorp, 1969 infesting *Alveolitella fecunda* (Lecompte). Trzemoszna, Holy Cross Mountains, Poland; Late Givetian. Longitudinal section. Specimen ZPAL T.25 TR034. B, *Helicosalpinx* sp. infesting *Alveolitella fecunda* (Lecompte). Trzemoszna, Holy Cross Mountains, Poland; Late Givetian. Longitudinal section. Specimen ZPAL T.25 TR 035. C–F, *Cheatosalpinx? plusquelleci* sp. n., syntypes; infesting *Pachyfavosites polonicus* Nowiński, holotype. Sowie Górki, Holy Cross Mountains, Poland; set G, Middle Frasnian. Specimen ZPAL T.25 T. XVIII 24/2; thin sections ZPAL T.25 SG 200–202: C, transverse section (thin arrow marks thin tube between walls, thick arrow shows large diameter tube penetrating corallite lumen); D, longitudinal section, arrow marks longitudinal section of parasite. E, transverse section; F, transverse section. All scale bars represent 500 μm.

Occurrence. Ordovician–Devonian. Worldwide distribution (see Tapanila 2004).

Helicosalpinx ichnosp.
Figure 28B

Material. A corallum of *Alveolitella fecunda* (one thin section, ZPAL T.25 TR 035), infested by two individuals; from the Givetian of Trzemoszna.

Description. Helicoidally curved tubes placed between the corallite walls. Within the 3-mm-long longitudinal section, four whorls are visible. The coiling is regular and tight, with radius larger than the tube diameter. The maximal diameter of tube is about 0.20 mm. Cross section unknown.

Remarks. The coiling in *Helicosalpinx* isp. is much more regular than in *H. asturiana.* It is possible that both specimens belong to one species, but the material is scarce and therefore it is difficult to assign them to one and same taxon; hence, at this stage, they have been treated as potential separate taxa.

The longitudinal section is nearly the same as figured by Hubmann (1991*a*, pl. 3, fig. 1), but its spiral shape has probably higher angle of coiling than those illustrated by Oekentorp (1969, pl. 15, fig. 1) or Tapanila (2004, fig. 3).

The position of infesting organisms within the host corallum

The individuals of *Chaetosalpinx? plusquelleci* isp. nov. occur, similar to *Ch. ferganensis* (see Zapalski 2007*a*, fig. 1), between the corallite walls, and in contrary to the latter species within the corallite lumina. Therefore, they both must have perforated the host's soft tissue. A strong modification of host skeletal geometry is visible, similar to that described by Oekentorp (1969) and Zapalski (2007*a*).

Similar to most specimens of *Chaetosalpinx*, the *Helicosalpinx* species described above occur between the walls. As the walls are relatively thick, it is difficult to see whether modifications of the skeletal organization occur. The host wall around the infesting organism is darkened, and it looks very similar to darkening of the wall in traces of the median line. It might have been caused by a different arrangement of crystals around the infesting organism, but scarcity of material causes that it has not been possible to make ultrathin sections in order to investigate microstructure.

The interaction between infesting organisms and host coral

The influence of *Chaetosalpinx? plusquelleci* isp. nov. described here on the morphology of host coral is very

similar to the other known species *C. ferganensis.* Their presence within the lumina of host corallites did cause competition for space between host and endosymbiont.

All *Helicosalpinx* reported up to now had visible influence on the morphology of neighbouring corallites (e.g. Oekentorp 1969; Stel 1976, fig. 7b; Hubmann 1991*a*, pl. 3 fig. 1; Tapanila 2004). In the specimens investigated here, such influence on skeletal morphology is not easy to observe. Tapanila (2004, pp. 105–106) in his analysis of Ordovician *Helicosalpinx* mentions 'minor disruption of normal septal growth' (observing that the overall growth was not disrupted). Such disruptions are the 'modifications in the skeletal organization of Favositida' mentioned by Oekentorp (1969). The lack of such influence on host skeleton may be caused by scarcity of material. Such an influence on the host anatomy can be (according to Zapalski 2007*a*) treated as Dawkins's (1982) 'extended phenotype', where the endobiont causes change in the coral anatomy. Thus, *Helicosalpinx* organisms had modified the host's phenotype in many cases. Moreover, these organisms are placed within the skeleton; such a placement was not possible without perforation of the soft tissue (Zapalski 2007*a*). Perforation of soft tissue is clearly negative effect on the host. The time of coexistence of these two organisms cannot be clearly estimated, as on investigated specimens growth bands are absent. Deep (several millimetres) embedding in host's skeleton suggests long coexistence.

The following signs: presumably long coexistence and perforation of the host's tissues can lead to the conclusion that *Chaetosalpinx? plusquelleci* isp. nov., *H.* cf. *asturiana* Oekentorp and *H.* isp., and *Chaetosalpinx ferganensis* Sokolov were rather parasites of tabulate corals than their commensals, as previously suggested, although their influence on host's anatomy is not that well visible on studied material as on previously published material.

STRATIGRAPHIC AND PALAEOGEOGRAPHIC DISTRIBUTION OF TABULATES IN THE KIELCE REGION

The palaeogeographic and stratigraphic distribution of Devonian tabulates from Poland was already studied by Nowiński (1992), but his observations were made on generic level; moreover, he did not analyse the distribution of tabulate corals within the Kielce Region but compared Kielce Region (*sensu stricto*) as a whole with the Kostomłoty Transitional Zone and the Silesia–Cracow Region.

Incomplete data such as the lack of Givetian faunas in the Northern Subregion or Frasnian faunas in the Kostomłoty Transitional Zone make comparisons very difficult; nevertheless, several patterns can be observed (data

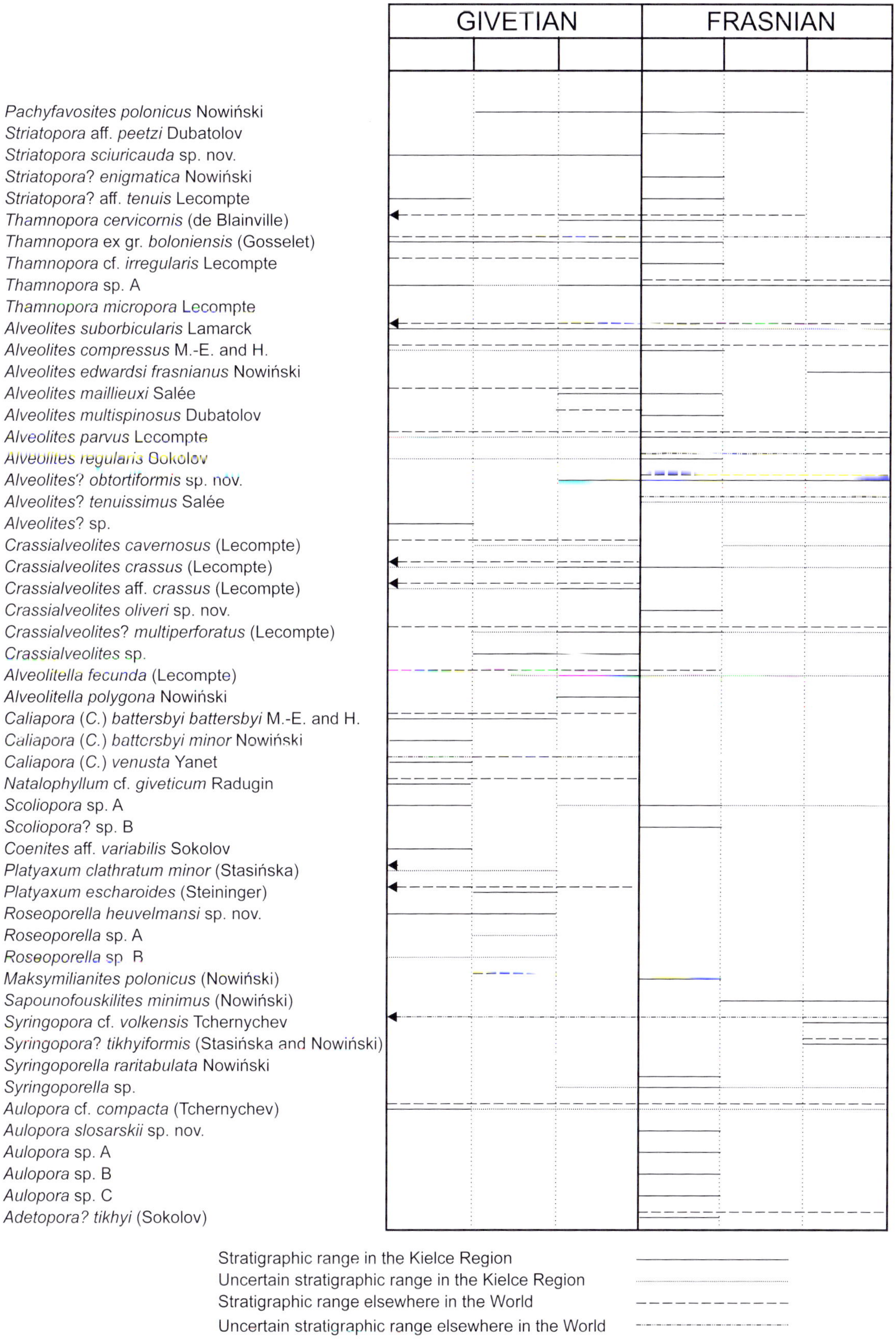

GIVETIAN	FRASNIAN

Pachyfavosites polonicus Nowiński
Striatopora aff. peetzi Dubatolov
Striatopora sciuricauda sp. nov.
Striatopora? enigmatica Nowiński
Striatopora? aff. tenuis Lecompte
Thamnopora cervicornis (de Blainville)
Thamnopora ex gr. boloniensis (Gosselet)
Thamnopora cf. irregularis Lecompte
Thamnopora sp. A
Thamnopora micropora Lecompte
Alveolites suborbicularis Lamarck
Alveolites compressus M.-E. and H.
Alveolites edwardsi frasnianus Nowiński
Alveolites maillieuxi Salée
Alveolites multispinosus Dubatolov
Alveolites parvus Lecompte
Alveolites regularis Sokolov
Alveolites? obtortiformis sp. nov.
Alveolites? tenuissimus Salée
Alveolites? sp.
Crassialveolites cavernosus (Lecompte)
Crassialveolites crassus (Lecompte)
Crassialveolites aff. crassus (Lecompte)
Crassialveolites oliveri sp. nov.
Crassialveolites? multiperforatus (Lecompte)
Crassialveolites sp.
Alveolitella fecunda (Lecompte)
Alveolitella polygona Nowiński
Caliapora (C.) battersbyi battersbyi M.-E. and H.
Caliapora (C.) battersbyi minor Nowiński
Caliapora (C.) venusta Yanet
Natalophyllum cf. giveticum Radugin
Scoliopora sp. A
Scoliopora? sp. B
Coenites aff. variabilis Sokolov
Platyaxum clathratum minor (Stasińska)
Platyaxum escharoides (Steininger)
Roseoporella heuvelmansi sp. nov.
Roseoporella sp. A
Roseoporella sp. B
Maksymilianites polonicus (Nowiński)
Sapounofouskilites minimus (Nowiński)
Syringopora cf. volkensis Tchernychev
Syringopora? tikhyiformis (Stasińska and Nowiński)
Syringoporella raritabulata Nowiński
Syringoporella sp.
Aulopora cf. compacta (Tchernychev)
Aulopora slosarskii sp. nov.
Aulopora sp. A
Aulopora sp. B
Aulopora sp. C
Adetopora? tikhyi (Sokolov)

Stratigraphic range in the Kielce Region
Uncertain stratigraphic range in the Kielce Region
Stratigraphic range elsewhere in the World
Uncertain stratigraphic range elsewhere in the World

FIG. 30. Stratigraphic distribution of tabulate corals in the Givetian-Frasnian of the Kielce Region.

TABLE 35. Distribution of tabulate corals from the Devonian of the southern region of the Holy Cross Mountains, Poland.

	KTZ		Northern Kielce SR							Central Kielce SR						Southern Kielce SR			
	Laskowa Q.	Czarnów	Kadzielnia	Szczukowskie G.	Grabina	Wietrznia	G. Cmentarna	Psie G.	Miedzianka	Marzysz	Sitkówka-Kostrzewa	Sowie G.	Bolechowice-Panek	Posłowice	Trzemoszna	Jaźwica/G.Łgawa	Kowala	Stokówka	Bilcza
Pachyfavosites polonicus Nowiński	A–B													1					
Striatopora sciuricauda sp. nov.		A								1		G							B
Striatopora aff. *peetzi* Dubatolov																	1		
?*Striatopora enigmatica* Nowiński																	A		
?*Striatopora* aff. *tenuis* Lecompte																	A–C		
Thamnopora ex gr. *boloniensis* (Gosselet)	1	A									1	B–C				H	A–C		
Thamnopora cervicornis (de Blainville)											1						1		
Thamnopora cf. *irregularis* Lecompte																H	A?		
Thamnopora micropora Lecompte	A?																C		
Thamnopora sp. A	?A–?B															1			
Alveolites suborbicularis Lamarck	A	B	A	1		1	1	1	1					1		I–K	A–C		
Alveolites compressus Milne-Edwards and Haime	1							1	1							J–K	A		
Alveolites edwardsi frasnianus Nowiński																1			
Alveolites maillieuxi Salée		C														I–K	A		
Alveolites multispinosus Dubatolov																	C		
Alveolites parvus Lecompte	1			1		1	1	1								J?–K?			
Alveolites regularis Sokolov	1		A									1				A			
Alveolites? obtortiformis sp. nov.									1						1				
Alveolites? tenuissimus Lecompte								1								1			
Alveolites sp.	1							1											
Crassialveolites crassus (Lecompte)		1				1	1	1				1		1					
Crassialveolites cavernosus (Lecompte)					C							1		1					
Crassialveolites? multiperforatus (Lecompte)						1	1					1		1				C	
Crassialveolites oliveri sp. nov.																			
Crassialveolites aff. *crassus* (Lecompte)	A?								1			1		1			C?		
Crassialveolites sp.		B										B							
Alveolitella fecunda (Lecompte)	B								1					1	1				
Alveolitella polygona Nowiński														B					
Caliapora (C.) battersbyi Milne-Edwards and Haime	A?	A																	

TABLE 35. (*Continued*)

	KTZ		Northern Kielce SR						Central Kielce SR							Southern Kielce SR			
	Laskowa Q.	Czarnów	Kadzielnia	Szczukowskie G.	Grabina	Wietrznia	G. Cmentarna	Psie G.	Miedzianka	Marzysz	Sitkówka-Kostrzewa	Sowie G.	Bolechowice-Panek	Posłowice	Trzemoszna	Jaźwica/G.Łgawa	Kowala	Stokówka	Bilcza
Caliapora (C.) battersbyi minor Nowiński		A																	
Caliapora (C.) venusta Yanet		A																	
Natalophyllum cf. *giveticum* Radugin		A																	
Scoliopora sp.		A	A					1			1					K?			
Scoliopora? sp. B	A?																A		
Coenites aff. *variabilis* Sokolov	A?																		
Platyaxum clathratum minus (Stasińska)	1																		
Platyaxum escharoides (Steininger)	1																		
Roseoporella heurelmansi sp. nov.	B																		
Roseoporella sp. A	B?																		
Roseoporella sp. B	A–B?																		
Maksymiliannites poloniensis (Nowiński)																			
Sapounofouskilites minimus (Nowiński)													1			R			
Syringopora cf. *volkensis* Tchernychev																R			
Syringopora? tikhyiformis (Stasińska and Nowiński)																R			
Syringoporella raritabulata Nowiński						A						G							
Syringoporella sp.												K							
Aulopora cf. *compacta* (Tchernychev)	A?					1									1				
Aulopora slosarskii sp. Nov.																	C		
Aulopora sp. A																	C?		
Aulopora sp. B																	A		
Aulopora sp. C																	A		
Adetopora? tikhyi (Sokolov)																1	A–C		

1 – species present in unspecified lithological set; lithological set numbers (A–R) after Racki (1992a). KTZ, Kostomłoty Transitional Zone.

concerning occurrence of tabulates in the Kielce Region is given in Table 35 and Figure 30):

1. Each Subregion, except for the Northern Subregion, has taxa restricted to the given region only, namely:
   a. Givetian of the Kostomłoty Transitional Zone: *Roseoporella heuvelmansi, Caliapora (C.) battersbyi minor*;
   b. Givetian of the Central Kielce Subregion: *Alveolitella polygona* and Givetian/Frasnian *Pachyfavosites polonicus*;
   c. Frasnian of the Central Kielce Subregion: *Syringoporella raritabulata*;
   d. Frasnian of the Southern Kielce Subregion: *Crassialveolites oliveri, Aulopora slosarskii* and *Alveolites edwardsi frasnianus*.
2. As already observed by Nowiński (1992), characteristic assemblage of *Alveolitella* occurs solely in the late Givetian of Central Kielce Subregion; presence of numerous *Alveolitella* in the Kostomłoty Transitional Zone reported by Nowiński (1992; *Alveolitella fecunda* Assemblage in his paper) is not confirmed by the present study;
3. Coenitids seem to be restricted to the Givetian of the Kostomłoty Transitional Zone. They were probably preferring more argillaceous sediments (they are numerous in the mixed facies of the Łysogóry Region, Stasińska 1958);
4. Members of the family Natalophyllidae do not occur in the Central Kielce Subregion; they are rare in other parts of Kielce Region;
5. The Southern Kielce Subregion has a very diverse assemblage of auloporids (*Aulopora slosarskii, Aulopora* sp. A, *Aulopora* sp. B, *Aulopora* sp. C and *Adetopora*? *tikhyi*), and a single species (*Aulopora* cf. *compacta*) occurs in other subregions. The auloporid assemblage from the Southern Subregion is smaller than that known from the Łysogóry Region (e.g. late Eifelian of Grzegorzowice–Skały section, six species, Zapalski 2003).
6. Syringoporids are generally restricted to Southern and Central Kielce Subregions (six species), and only single specimen of *Syringella* sp. was found in Northern Subregion.
7. Species rare in the Kielce Region (e.g. *Maksymilianites polonicus, Alveolites multispinosus, Alveolites*? *obtortiformis*) seem to be rare also elsewhere in the world, while species common in the Kielce Region (e.g. *Alveolites parvus, A. suborbicularis, Thamnopora* ex gr. *boloniensis*) seem to be ubiquitous and common also in other areas.

The Givetian tabulates from the Kielce Region are dominated (by number of species) by the Alveolitidae (42 per cent), and second and third large groups are the families Pachyporidae and Coenitidae (18 per cent each; see Fig. 30). Such a situation is different from that known from Ardennes, where pachyporids are dominating, and the Alveolitidae are second largest group (Zapalski *et al.* 2007*b*). In contrast, the Frasnian composition is similar between these two regions. In the Frasnian of Kielce Region, the Alveolitidae constitute 39 per cent of all tabulates and the Pachyporidae 22 per cent, and the dominance of Alveolitidae is less marked. In the Frasnian of Ardennes, members of Alveolitidae are also visibly dominating, while Pachyporidae are less abundant (Zapalski *et al.* 2007*b*). Therefore, the turnover in faunal domination between pachyporids and alveolitids known from the Givetian – Frasnian of Ardennes (Givetian: dominating pachyporids; Frasnian: dominating alveolitids) does not occur in the Kielce Region; alveolitids dominate during the investigated periods, always constituting about 40 per cent of tabulate faunas.

CONCLUSIONS

1. In the Devonian of the Kielce Region (including the Kostomłoty Transitional Zone) occur 52 taxa of tabulate corals. They belong to three orders: Favositida (40 species and subspecies), Syringoporida and Auloporida (six species each).
2. Out of that number, five species are new; moreover, a new genus was erected: a syringoporid with thick walls, thick-walled axial canal and dissepimental tissue – *Sapounofouskilites* gen. nov.
3. In taxonomy of alveolitids, the most useful seem to be (1) corallite (lumen) diameters and (2) pore diameter. Double wall thickness and spacing of tabulae are less helpful. Last two parameters may have very high intracolonial variation. Intracolonial variation in alveolitids and coenitids is higher than in heliolitids.
4. In taxonomy of coenitids, likewise in alveolitids, corallite (lumen) diameters and pore diameters seem to be most useful. This group requires, however, further research.
5. Organisms infesting tabulate corals, assigned to *Chaetosalpinx*? *plusquelleci* isp. nov., *Helicosalpinx* cf. *asturiana* Oekentorp and *H.* sp., are rather tabulate parasites than their commensals, as previously postulated. Representatives of the genus *Helicosalpinx* are recognized in Poland for the first time.
6. Givetian and Frasnian tabulates are dominated by representatives of Alveolitidae. The Givetian composition of the tabulate faunas is different from the one known from Franco–Belgian Ardennes, where Pachyporidae were dominating during the Givetian. Frasnian domination of alveolitids in the Kielce Region is similar to the situation known from Ardennes.

Acknowledgements. This monograph derived from my Ph. D. thesis under the direction of Professors Bruno Mistiaen (Lille, France) and Jerzy Trammer (Warsaw), to whom I am very indebted for tutorship, discussions and help. Thesis referees, Esperanza Fernández-Martínez (León, Spain), Urszula Radwańska and Ewa Roniewicz (both from Warsaw) gave numerous precious remarks. Also journal referees, Andreas May (Madrid) and Jindřich Hladil (Prague) gave important remarks that helped to improve this work.

I am also deeply grateful to Aleksander Nowiński (Warsaw) and Yves Plusquellec (Brest, France) for their help and consultations. My professors and colleagues from Lille are thanked for critical discussions: Denise Brice, Catherine Crônier, Benoît L. M. Hubert, Bruno Milhau and Jean-Pierre Nicollin. Separate thanks are due to Pascal Deville, for preparation of thin sections. Marie Coen-Aubert (Bruxelles) kindly gave access to the collections of M. Lecompte housed at the Musée de l'Institut royal des Sciences naturelles, Bruxelles. Andreas May discussed with me the coenitid taxonomy, and Shuji Niko (Hiroshima, Japan) benevolently provided some hard-to-find publications. Several of my English-speaking friends and colleagues – John Brenner (Wokingham, UK), Euan N. K. Clarkson (Edinburgh, UK), Paul Copper (Sudbury, Canada), Steve Kershaw (London), Gregg Webb (Brisbane, Australia) and Graham Young (Winnipeg, Canada) – kindly improved the English of several chapters of this memoir. My friend Adam T. Halamski (Warsaw) was assisting me in the field and discussing the manuscript. I wish to express my deep gratitude to all of them. Last but not least, I would like to express my gratitude to my wife Ewa, for her endless patience and time devoted for cross-checking my references.

The financial support of the Institut Supèrieur d'Agriculture (Lille), the French Government (scholarship for foreign Ph D students) and Foundation for Polish Science ('Start' scholarship for 2007–2008) is deeply appreciated.

REFERENCES

ALKHOVIK, T. S. 1985. On the systematic position and phylogenetic relationships of the genus Scoliopora (Favositida). *Palaeontologicheski Zhurnal*, **1985** (3), 20–26. [In Russian]

BAE, B.-Y., ELIAS, R. J. and LEE, D.-J. 2006. Morphometrics of *Catenipora* (Tabulata; Upper Ordovician; southern Manitoba, Canada). *Journal of Paleontology*, **80**, 889–901.

—— —— —— 2008. Morphometrics of *Manipora* (Tabulata; Upper Ordovician; southern Manitoba, Canada). *Journal of Paleontology*, **82**, 78–90.

BAIKUCHKAROV, A. G. 1992. On the variability in the species *Crassialveolites multiperforatus* (Salée) (Tabulata). *Paleontologicheskii Zhurnal*, **26** (1), 104–108.

BERKOWSKI, B. 1991. A blind phacopid trilobite from the Famennian of the Holy Cross Mountains. *Acta Palaeontologica Polonica*, **36**, 255–264.

—— 2002. Famennian Rugosa and Heterocorallia from Southern Poland. *Palaeontologia Polonica*, **61**, 1–87.

BERTLING, M., BRADDY, S. J., BROMLEY, R. G., DEMATHIEU, G. R., GENISE, J., MIKULAS, R.,

NIELSEN, J. K., NIELSEN, K. S. S., RINDSBERG, A. K., SCHLIRF, M. and UCHMAN, A. 2006. Names for trace fossils: a uniform approach. *Lethaia*, **39**, 265–286.

BERTRAND, M., COEN-AUBERT, M., DUMOULIN, V., PRÉAT, A. and TOURNEUR, F. 1993. Sédimentologie et paléoécologie de l'Emsien supérieur et d'Eifelien inférieur des régions de Couvin et de Villers-la-Tour (bord sud du Synclinorium de Dinant, Belgique). *Neues Jahrbuch für Geologie und Paläontologie – Abhandlungen*, **188**, 177–211.

BEST, M. B., BOEKSCHOTEN, G. J. and OOSTERBAAN, A. 1984. Species concept and ecomorph variation in living and fossil scleractinia. *Palaeontographica Americana*, **54**, 70–79.

BIERNAT, G. 1953. O trzech nowych brachiopodach z tzw. wapienia stringocefalowego Gór Świętokrzyskich. *Acta Geologica Polonica*, **3**, 299–324.

—— 1959. Middle Devonian Orthoidea of the Holy Cross Mountains and their ontogeny. *Palaeontologia Polonica*, **10**, 1–78.

—— 1964. Middle Devonian Atrypacea (Brachiopoda) from the Holy Cross Mountains, Poland. *Acta Palaeontologica Polonica*, **9**, 277–336.

—— 1966. Middle Devonian brachiopods of the Bodzentyn syncline (Holy Cross Mountains, Poland). *Palaeontologia Polonica*, **17**, 1–162.

BIRENHEIDE, R. 1985. Chaetetida und tabulate Korallen des Devon. *Leitfossilien*, **3**, 1–158.

—— 1998. Rugose und tabulate Korallen aus der Bohrung Viersen 1001. *Fortschritte in der Geologie von Rheinland und Westfalen*, **37**, 161–213.

BLAINVILLE, de H. 1830. Zoophytes. *Dictionnaire des Sciences Naturelles*, **60**, 1–546.

BOULVAIN, F., COEN-AUBERT, M. and TOURNEUR, F. 1987. Sedimentologie et coraux du bioherme de marbre rouge frasnien ("F2j") de Tapoumont (Massif de Philippeville, Belgique). *Annales de la Société Géologique de Belgique*, **110**, 225–240.

BRICE, D., BIGEY, F., MISTIAEN, B., PONCET, J. and ROHART, J.-C. 1975. Les organisms constructeurs (Algues, Stromatopores, Rugeux, Tabulés, Bryozoaires) dans le Dévonien de Ferques (Boulonnais-France). Associations – Répartition stratigraphique. *Mémoires BRGM*, **89**, 136–151.

BROOD, K. 1970. The systematic position of *Coenites* Eichwald. *Geologiska Föreningens i Stockholm Förhandlingar*, **92**, 469–480.

BRÜHL, D. 1996. Die Gattungen *Alveolites* Lamarck 1801 und *Squameoalveolites* MIRONOVA 1969 (Anthozoa, Tabulata) im Unteren Mittel-Devon (Eifelium) der Dollendorfer Mulde/Eifel (Rheinisches Schiefergebirge). *Senckenbergiana lethaea*, **76**, 1–51.

—— 1999. Stratigraphie, Fazies und Tabulaten-Fauna des oberen Eifelium (Mittel-Devon) der Dollendorfer Mulde/Eifel (Rheinisches Schiefergebirge). *Kölner Forum für Geologie und Paläontologie*, **4**, 1–155.

—— and POHLER, S. 1999. Tabulate Corals from the Moore Creek Limestone (Middle Devonian: Late Eifelian–Early Givetian) in the Tamworth Belt (New South Wales, Australia). *Abhandlungen der Geologischen Bundesanstalt*, **54**, 275–293.

BUDD, A. F. 1993. Variation within and among morphospecies of *Montastraea*. *Courier Forschungsinstitut Senckenberg*, **164**, 241–254.

—— and STOLARSKI, J. 2009. Searching for new morphological characters in scleractinian reef corals: comparison of septal teeth and granules between Atlantic and Pacific Mussidae. *Acta Zoologica*, **90**, 142–165.

BYRA, H. 1983. Revision der von Cl. Schlüter (1880–1889) beschriebenen Chaetetida und Tabulata aus dem Rheinischen Devon. *Courier Forschungsinstitut Senckenberg*, **59**, 1–78.

CHATTERTON, B. D. E., COPPER, P., DIXON, O. A. and GIBB, S. 2008. Soft polyps with spicular sclerites in Silurian favositid corals from Anticosti Island, E. Canada, and Silurian heliolitids from the Canadian arctic. *Palaeontology*, **51**, 173–198.

CHLUPÁČ, I. 1992. Trilobites from the Givetian and Frasnian of the Holy Cross Mountains. *Acta Palaeontologica Polonica*, **37**, 395–406.

CLARKE, J. M. 1908. The beginnings of dependent life. *New York State Museum Bulletin*, **121**, 146–196.

COEN-AUBERT, M., DEJONGHE, L., CNUDDE, C. and TOURNEUR, F. 1985. Etude stratigraphique, sedimentologique et geochimique de trois sondages effectues a Membach (Massif de la Vesdre). *Ministere des Affaires Economiques, Professional Paper*, **223**, 3–57.

COMBES, C. 2001. *Parasitism: the Ecology and Evolution Intimate Interactions*. University of Chicago Press, Chicago, 552 pp.

COPPER, P. 1985. Fossilized polyps in 430-Myr-old *Favosites* corals. *Nature*, **316**, 142–144.

CZARNOCKI, J. 1950. Geologia regionu Łysogórskiego w związku z zagadnieniem rud żelaza w Rudkach. *Prace Państwowego Instytutu Geologicznego*, **6a**, 1–75.

—— 1989. Klimenie Gór Świętokrzyskich. *Prace Państwowego Instytutu Geologicznego*, **127**, 1–91.

DANA, J. D. 1846. *Structure and classification of zoophytes: U. S. exploring expedition during years 1838, 1839, 1840, 1841, 1842 under the command of Charles Wilkes, U. S. N. Volume 7*. Lea & Blanchard, Philadelphia, x +, 740 pp., 61 pls.

DAVIS, W. J. 1887. *Kentucky Fossil Corals, a Monograph on the Fossil Corals of the Silurian and Devonian Rocks of Kentucky, pt. 2.* Kentucky Geological Survey, Frankfort, 139 pp., 4 pls.

DAWKINS, R. 1982. *The Extended Phenotype*. W. H. Freeman and Company, Oxford, 307 pp.

DENG ZHAN-QIU 1979. Middle Devonian Tabulate corals and Chaetetids from Dushan, Southern Guizhou. *Acta Palaeontologica Sinica*, **18**, 151–160.

DIXON, O. A. 2010. Fossilized polyp remains in Silurian Heliolites (Anthozoa, Tabulata) from Nunavut, Arctic Canada. *Lethaia*, **43**, 60–72.

DUBATOLOV, V. N. 1956. Tabulate corals and heliolitids if NE Prisalair. *Ezhegodnik Vsesoyuznogo Paleontologicheskago Obshchcestva*, **15**, 83–114. [In Russian].

—— 1959. Tabulates, heliolitids and chaetetids of the Silurian and Devonian from the Kuznetsk Basin. *VNIGRI*, **139**, 1–292. [In Russian]

—— 1963. *Late Silurian and Devonian Tabulates, heliolitids and chaetetids from the Kuznetsk Basin*. AN SSSR, Sibirskoe otdele-nie, Institut Gieologii i Gieofiziki, Izdatielstvo AN SSSR, Moskva, pp. 1–194. [In Russian]

—— 1969. Tabulates and biostratigraphy of the early Devonian of the NE Soviet Union. *Transaction of the Institute of Geology and Geophysics*, **70**, 1–175.

—— and IVANOVSKI, A. B. 1977. Index of tabulate genera. *Trudy Instituta Gieologii i Gieofiziki*, **336**, 3–155. [In Russian]

—— and SPASSKI, N. YA. 1971. Devonian corals of the Dzhunghara-Balkhash Province. *Trudy Instituta Gieologii i Gieofiziki*, **74**, 1–131. [In Russian]

—— TCHEKHOVICH, V. D. and YANET, F. E. 1968. Tabulata of the Silurian–Devonian border beds in Altai-Saian Upper Province and in Ural Mts. 5–109. *In* IVANOVSKI, A. B. (ed.) *Corals of the Silurian–Devonian border beds in Altai-Saian Upper Province and in Ural Mts.* Nauka, Moskva, 109 pp.

DUNCAN, P. M. 1872. Third report on the British fossil corals. 116–137. Report of the 41st Meeting of the British Association for the Advancement of Science. John Murray, London, 208 pp.

DZIK, J. 2002. Emergence and collapse of the Frasnian conodont and ammonoid communities in the Holy Cross Mountains, Poland. *Acta Palaeontologica Polonica*, **47**, 565–650.

—— 2006. The Famennian "Golden Age" of conodonts and ammonoids in the Polish part of the Variscan sea. *Palaeontologia Polonica*, **63**, 1–360.

EHRENBERG, C. G. 1834. Beiträge zur physiologischen Kenntnis der Corallenthiere im allgemeinen, und besonders der Rothen Meeres, nebst einem Versuche zur physiologischen Systematik dersleben. *Kaiserliche Akademie der Wissenschaften, Physik.-Mathematik., Abhandlungen*, **1834**, 225–380.

EICHWALD, C. E. von. 1829. *Zoologia Specialis quam expositis animalibus tum vivis tum fossilibus potissimum rossiae in universum, et poloniae in specie, in usum lectionum.* Józef Zawadzki, Vilnius, Vol. 1, vi + 314, 5 pls.

—— 1860. *Lethaea Rossica ou Paléontologie de la Russie. Premier volume.* E. Schweizerbart, Stuttgart, 681 pp., 59 pls.

—— 1861. *Palaeontology of Russia. Older period. II.* Golike, St. Petersburg, 521 pp.

ERMAKOVA, K. A. 1958. Auloporids of the Central Devonian Field. *Trudy VNIGRI*, **9**, 28–39. [In Russian].

FEDOROWSKI, J. 1989. Intraspecific variation in Carboniferous and Permian Rugosa. *Memoir of the Association of Australasian Palaeontologists*, **8**, 7–12.

FENTON, M. A. and FENTON, C. L. 1937. Aulopora: a Form-Genus of Tabulate Corals and Bryozoans. *American Midland Naturalist*, **18**, 109–115.

FERNÁNDEZ-MARTÍNEZ, E. and MISTIAEN, B. 2003. *Alveolites parvus*, tabulate coral from Upper Devonian of Iran. *Annales de la Société Géologique du Nord, Series 2*, **10**, 261–273.

—— and TOURNEUR, F. 1993. El genero *Caliapora* (Tabulata) en el Devónico de la Cordilliera Cantabrica (NW de España). *Revista Española de Paleontologia*, **No. Extraordinario**, 58–70.

FLÜGEL, H. W. 1976. Ein Spongienmodell für die Favositidae. *Lethaia*, **9**, 405–419.

FONTAINE, H. 1954. Etude et revision des tabulés et helioli-
tides du Devonien d'Indochine et du Yunnan. *Archives Geologiques du Vietnam*, **2**, 1–86.

—— 1959. Les Madréporaires paleozoïques du Viet-Nam, du
Cambodge et du Laos. Unpublished PhD Thesis, Faculté des
Sciences de l'Universite de Paris, Paris, 430 pp.

—— 1961. Les Madréporaires paleozoïques du Viet-Nam, du
Laos et du Cambodge. *Archives Geologiques du Vietnam*, **5**,
1–276.

FOOTE, M. and MILLER, A. I. 2007. *Principles of Paleontology*, Third Edition. W. H. Freeman and Co., New York, xv +
354 pp.

FOSTER, A. B. 1984. The species concept in fossil hermatypic
corals: a statistical approach. *Paleontographica Americana*, **54**,
58–69.

FROMENTEL, de E. 1861. *Introduction a l'étude des polypiers
fossiles*. F. Savy, Paris, 357 pp.

FUKAMI, H., BUDD, A. F., PAULAY, G., SOLÉ-CAVA,
A., CHEN, C. A., IWAO, K. and KNOWLTON, N. 2004.
Conventional taxonomy obscures deep divergence between
Pacific and Atlantic corals. *Nature*, **427**, 832–835.

GERTH, H. 1921. Coelenterata. Anthozoa. *In*: MARTIN, K.
(ed.) *Die Fossilien von Java auf Grund einer Sammlung von Dr.
R. D. M. Verbeek und von Anderen. Sammlungen des Geologische Reichsmuseums in Leiden*, **1**, 387–445.

GOLDFUSS, A. 1826. *Petrefacta Germaniae, Band I, Teil I*.
Arnz, Düsseldorf, pp. 1–76, 25 pls.

—— 1829. *Petrefacta Germaniae, Band I, Teil II*. Arnz, Düsseldorf, pp. 77–164, 24 pls.

GONZÁLEZ-FORERO, M. 2009. Removing ambiguity from
the biological species concept. *Journal of Theoretical Biology*,
256, 76–80.

GOSSELET, J. 1877. Le calcaire dévonien supérieur dans le
N.–E. de l'arrondisment d'Avesnes. *Annales de la Société
Géologique du Nord*, **4**, 238–272.

—— 1880. *Esquisse géologique du Nord de la France et contrées
voisines. 1er fascicule, terrains primaires*. Imprimerie Six-Hore-
mans, Lille, 343 pp., 28 pls.

GÜRICH, G. 1896. Das Palaeozoikum im Polnischen
Mittelgebirge. *Verhandlungen der Russischen Mineralogischen
Kaiserlichen Gesselschaft zu St.Petersburg II*, **32**, 1–539.

HAJŁASZ, B. 1992. Tentaculites from the Givetian and
Frasnian of the Holy Cross Mountains. *Acta Palaeontologica
Polonica*, **37**, 385–394.

HALAMSKI, A. T. and BALIŃSKI, A. 2009. Latest Famennian brachiopods from Kowala, Holy Cross Mountains,
Poland. *Acta Palaeontologica Polonica*, **54**, 289–306.

HALL, J. 1851. New genera of fossil corals from the report by
James Hall, on the Palaeontology of New York. *American
Journal of Science, Series II*, **11**, 398–401.

HARTMAN, W. D. and GOREAU, T. F. 1975. A Pacific tabulate sponge, living representative of a new order of sclerosponges. *Postilla*, **167**, 1–21.

HEY, J. 2006. On the failure of modern species concepts. *Trends
in Ecology and Evolution*, **21**, 447–450.

HILL, D. 1981. Tabulata. F430–F762. *In* MOORE, R. C. and
TEICHERT, C. (eds). *Treatise on invertebrate paleontology.
Part F, Coelenterata. Supplement 1 (2)*. The Geological Society
of America, Boulder, and The University of Kansas, Lawrence,
388 pp.

—— and BUTLER, A. J. 1936. *Cymatelasma*, a new genus of
Silurian rugose corals. *Geological Magazine*, **73**, 516–
527.

HLADIL, J. 1974. Tabulate corals from the Paleozoic basement
of the Carpathian foredeep (borehole Nitkovice-2). *Věstnik
Ústředniho Ústavu Geologického*, **49**, 219–222.

—— 1979. Reefal fauna from the Devonian limestones at Malho-
stovice (eastern border of the Boskovice furrow). *Věstnik
Ústředniho Ústavu Geologického*, **53**, 179–183.

—— 1980. K otázce stanovování hranice givet/frasn v devon-
ském vápencovém komplexu na svazích českeho masívu.
Zemný Plyn a Nafta, **25**, 25–32.

—— 1981a. *Alveolites* corals from the Middle and Upper Devonian of the Moravian Karst (Anthozoa, Tabulata). *Časopis
Moravského Muzea*, **66**, 25–32.

—— 1981b. The genus *Caliapora* Schlüter (tabulate corals) from
the Devonian of Moravia. *Věstnik Ústředniho Ústavu Geologického*, **56**, 157–168.

—— 1984a. Tabulate corals of the genus *Thamnopora* Steininger
from the Devonian of Moravia. *Věstnik Ústředniho Ústavu
Geologického*, **59**, 29–39.

—— 1984b. Tabulátní koralí z vrtu NP–824 Ostravice. *Acta Uni-
versitatis Carolinae-Geologica*, **1984**, 251–259.

—— 1987. The Lower Fammenian tabulate corals from the
southern Moravia. *Věstnik Ústředního ústavu geologického*, **62**,
41–46.

—— 1989a. Function morphology of Alveolitinae and its depen-
dence on the Kellwasser and other events (Tabulata, M. to U.
Devonian, Moravia, ČSSR). *Newsletter on Stratigraphy*, **21**,
25–37.

—— 1989b. Branched tabulate corals from the Koněprusy reef
(Pragian, Lower Devonian, Barrandian). *Věstnik Ústředního
ústavu geologického*, **64**, 221–230.

HOEKSEMA, B. W. 1993. Phenotypic corallum variability in
Recent mobile reef corals. *Courier Forschungsinstitut Senckenberg*, **164**, 263–272.

HÖFLING, R. 1989. Substrate-induced morphotypes and
intraspecific variability in Upper Cretaceous scleractinians of
the eastern Alps (West Germany, Austria). *Memoir of the Association of Australasian Palaeontologists*, **8**, 51–60.

HOWELL, B. F. 1962. Worms. W144–W177. *In* MOORE, R.
C. (ed.). *Treatise on Invertebrate Paleontology. Part W. Miscel-
lanea*. Geological Society of America and University of Kansas
Press, Lawrence, 259 pp.

HUBERT, B. L. M., ZAPALSKI, M. K., NICOLLIN, J.-P.,
MISTIAEN, B. and BRICE, D. 2007. Selected benthic faunas from the Devonian of the Ardennes: an estimation of palaeobiodiversity. *Acta Geologica Polonica*, **57**, 223–262.

HUBMANN, B. 1991a. Alveolitidae, Heliolitidae und *Helicosalpinx* aus den Barrandeikalken (Eifelium) des Grazer Devons.
Jahrbuch der Geologischen Bundesanstalt, **134**, 37–51.

—— 1991b. Halysitidae aus dem tiefen Silur E-Irans (Niur Formation). *Jahrbuch der Geologischen Bundesanstalt*, **134**, 711–733.

—— 1992. Variabilitätsuntersuchungen an *Catenipora* Lamarck
(Zoantharia, Tabulata). *Neues Jahrbuch für Geologie und Paläontologie – Monatshefte*, **1992**, 279–291.

HURCEWICZ, H. 1992. Middle and Late Devonia sponge spicules of the Holy Cross Mountains and Silesian Upland. *Acta Palaeontologica Polonica*, **37**, 291–296.

IMBRIE, J. 1956. Biometrical methods in the study of invertebrate fossils. *Bulletin of the American Museum of Natural History*, **108**, 211–252.

IVEN, C. 1980. Alveolitiden und Heliolitiden aus dem Mittel- und Oberdevon des Bergischen Landes (Rheinisches Schiefergebierge). *Palaeontographica Abteilung A*, **167**, 121–179.

—— MISTIAEN, B. and TOURNEUR, F. 1997. New data on the morphology and microstructure of the genus *Caliapora* Schlüter, 1889 (Tabulata, Middle Devonian): systematic implications and reflections on the function of squamulae. *Boletín de la Real Sociedad Española de Historia Natural*, **91**, 261–271.

JACKSON, J. B. C. and CHEETHAM, A. H. 1990. Evolutionary significance of morphospecies: a test with cheilostome Bryozoa. *Science*, **248**, 579–583.

JOHNSON, J. G., KLAPPER, G. and SANDBERG, C. A. 1985. Devonian eustatic fluctuations in Euramerica. *Bulletin of the Geological Society of America*, **96**, 567–587.

KAMIYA, T. and NIKO, S. 1998. A Late Silurian tabulate coral Planocoenites from the Hitoegane Formation, Gifu Prefecture, Japan. *[Name of Journal in Japanese]*, **47** (2), 67–70. [In Japanese, with an English Abstract].

KATO, M. 1963. Fine skeletal structures in Rugosa. *Hokkaido University, Journal of the Faculty of Sciences, Series 4*, **11**, 571–630.

KAŹMIERCZAK, J. 1971. Morphogenesis and systematics of the Devonian Stromatoporoidea from the Holy Cross Mts., Poland. *Palaeontologia Polonica*, **26**, 1–150.

—— 1984. Favositid Tabulates: evidence for Poriferan Affinity. *Science*, **225**, 835–837.

—— 1989. Halysitid tabulates: sponges in corals' clothing. *Lethaia*, **22**, 195–205.

—— 1993. Sclerite-bearing alveolitid favositids from the Devonian of Central Poland. *Paläontologische Zeitschrift*, **67**, 27–44.

—— 1994. Confirmation of the poriferan status of favositid tabulates. *Acta Palaeontologica Polonica*, **39**, 233–245.

KELUS, A. 1939. Devonische Brachiopoden und Korallen der Umgebung von Pelcza in Volhinien. *Biuletyn Państwowego Instytutu Geologicznego*, **8**, 1–51.

KETTNER, R. 1934. Paleontologické studie z Čelechovického Devonu, Čast 5: o některých Alcyonariich. *Časopis Vlasteneckého Muzea spoklu Olomuckého*, **47** (175–176), 1–15.

KHAIZNIKOVA, K. B. 1975. *Biostratigraphy and tabulate corals of the Devonian of Sette-Daban Ridge (southern part)*. Nauka, Novosibirsk, 112 pp. [In Russian].

KIELAN, Z. 1954. Les Trilobites mésodévoniens des Monts de Sainte-Croix. *Palaeontologia Polonica*, **6**, 1–50.

KIEPURA, M. 1965. Devonian bryozoans of the Holy Cross Mountains, Poland; Part I, Ctenostomata. *Acta Palaeontologica Polonica*, **10**, 11–48.

—— 1973. Devonian Bryozoans of the Holy Cross Mountains, Poland. Part 2. Cyclostomata and Cystoporata. *Acta Palaeontologica Polonica*, **18**, 323–400.

KIRKPATRICK, R. 1911. On *Merlia normani*, a Sponge with a Siliceous and Calcareous Skeleton. *Quarterly Journal of Microscopical Sciences*, **56**, 657–702.

—— 1912a. *Merlia normani* and its Relation to certain Palaeozoic Fossils. *Nature*, **89**, 502–503.

—— 1912b. On the nature of Stromatoporoids. *Nature*, **89**, 607.

KITCHER, P. 1984. Species. *Philosophy of Science*, **51**, 308–333.

KLAAMANN, E. 1959. On the tabulate faunas of Yuuruskyi and Tamsaluskyi horizonts. *Eesti NSV Teaduste Akademiaa Toimetised. Techniliste ja Füüsikaliss-Matemaatiliste Teaduste Seriaa*, **8**, 256–269. [In Russian].

KOKSCHARSKAYA, K. B. 1968. New Givetian alveolitids (Tabulata) of the Sette-Daban Ridge in Yakutia. *Paleontologicheskii Zhurnal*, **1968** (2), 21–25. [In Russian].

KRAEVSKAYA, L. N. 1955. Subclass Tabulata and the group Heliolitida. 191–206. *In* KHALFIN, L. L. (ed.). *Atlas of the leading forms of the fossil invertebrate faunas and floras of the Western Siberia*. Vol. 1. Gosgeoltekhnizdat, Moscow, 319 pp. [In Russian].

KRAWCZYŃSKI, W. 2002. Frasnian gastropod synecology and bio–events in the Dyminy reef complex of the Holy Cross Mountains, Poland. *Acta Palaeontologica Polonica*, **47**, 267–288.

—— 2006. Gastropod succession across the Early–Middle Frasnian transition in the Holy Cross Mountains, southern Poland. *Acta Palaeontologica Polonica*, **51**, 679–693.

KULICKA, R. and NOWIŃSKI, A. 1983. The Devonian Tabulata of the southern part of the Świetokrzyskie (Holy Cross) Mts., Poland. *Acta Palaeontologica Polonica*, **28**, 467–490.

LAFUSTE, J.-G. 1958a. Sur la microstructure des éléments horizontaux ches *Heliolites* Dana. *Comptes Rendus des Séances de l'Academie des Sciences*, **246**, 443–446.

—— 1958b. Note on the Structure and Microstructure of *Thecia swinderniana* (Goldf.). *Geological Magazine*, **95**, 409–414.

—— 1959. Sur la microstructure du genre *Striatopora* Hall, 1851. *Travaux du Laboratoire de Géologie de la Faculté des Sciences de l'Université de Paris. Annales Hebert et Haug*, **9**, 85–88.

—— 1963. Note on the microstructure of the Permian tabulate coral *Bayhaium*. *Journal of Paleontology*, **37**, 1127–1128.

—— 1970. Lames ultra-minces à faces polies. Procédé et application à la microstructure des Madréporaires fossiles. *Comptes Rendus de l'Academie des Sciences Séries D, Paris*, **270**, 679–681.

—— 1978. Modalites de passage des lamelles aux fibres dans la muraille de Tabules (Micheliniidae) du Devonien et du Permien. *Geobios*, **11**, 405–408.

—— 1979a. Microstructure de *Cladochonus* McCoy, 1847 (Tabulata, Carbonifère). *Geobios*, **12**, 353–363.

—— 1979b. Microstructure de type "desmidoïde" chez Lyopora Nicholson & Etheridge, 1878 (Tabulata, Ordovicien). *Comptes Rendus de l'Academie des Sciences, Séries D, Paris*, **289**, 719–722.

—— 1980. Sections ultra-minces de figures de corrosion à l'eau oxygenée. Procede et application aux lamelles et micro-lamelles des Tabulata. *Geobios*, **13**, 929–933.

—— 1983. Passage des microlamelles aux fibres dans le squelette d'un tabulé "michelinimorphe" du Viséen du Sahara Algérien. *Geobios*, **16**, 755–761.

—— 1984. Microstructure of Planalveolites Lang & Smith, 1939 (Tabulata, Ordovician). *Palaeontographica Americana*, **54**, 485–488.

—— 1986. Polymorphisme des fibres du sclérenchyme chez les Tabulés (Cnidaria, Paléozoïque). *Comptes Rendus de l'Academie des Sciences Séries II, Paris*, **302**, 761–763.

—— and DEBRENNE, F. 1970. Observation en lames ultraminces de la structure d'Archéocyathes. *C. R. Sommaire des Séances de la Sociéte Géologique de France*, **6**, 224–225.

—— and PLUSQUELLEC, Y. 1985. Structure et microstructure de quelques Micheliniidae et Michelinimorphes (Tabulata paleozoïques). *Bulletin du Museum National d'Histoire Naturelle, Paris*, **7** (4)C, 13–63.

—— and TOURNEUR, F. 1991. Microstructure du genre Acaciapora Moore & Jeffords, 1945 (Tabulata, Pennsylwanien de l'Oklahoma, USA). *Geologica et Palaeontologica*, **25**, 99–109.

LAMARCHE, J., MANSY, J. L., BERGERAT, F., AVERBUCH, O., HAKENBERG, M., LEWANDOWSKI, M., STUPNICKA, E., SWIDROWSKA, J., WAJSPRYCH, B. and WIECZOREK, J. 1999. Variscan tectonics in the Holy Cross Mountains (Poland) and of structural inheritance during Alpine tectonics. *Tectonophysics*, **313**, 171–186.

LAMARCK, J. B. 1801. *Systême des animaux sans vertèbres ou Tableau général des classes, des ordres et des genres de ces animaux*. Libraire Deterville, Paris, VIII + 432 pp.

LANG, J. C. 1984. Whatever works: the variable importance of skeletal and of non-skeletal characters in scleractinian taxonomy. *Palaeontographica Americana*, **54**, 18–44.

LANG, W. D. and SMITH, S. 1939. Some new generic names for Palaeozoic corals. *Annals and Magazine of Natural History, Series 11*, **3**, 152–156.

—— —— and THOMAS, H. D. 1940. *Index of Palaeozoic coral genera*. British Museum (Natural History), London, 231 pp.

LECOMPTE, M. 1933. Le genre *Alveolites* Lamarck dans le Dévonien moyen et supérieur de l'Ardenne. *Mémoires de l'Institut Royal des Sciences Naturelles de Belgique*, **55**, 1–50.

—— 1936. Révision des tabulés dévoniens décrits par GOLDFUSS. *Mémoires de l'Institut Royal de Sciences Naturelles de Belgique*, **75**, 1–112.

—— 1939. Les tabulés du Dévonien moyen et supérieur du bord sud du Bassin de Dinant. *Mémoires de l'Institut Royal des Sciences Naturelles de Belgique*, **90**, 1–229.

LEE, D. J. and NOBLE, J. P. A. 1988. Evaluation of corallite size as a criterion for species discrimination in favositids. *Journal of Paleontology*, **62**, 32–40.

LELESHUS, V. L. 1972. *Microalveolites* n. g. eine tabulate Koralle aus dem Unterdevon des Zeravšan-Gebirges (Tadžikistan). *Neues Jahrbuch für Geologie und Paläontologie, Monatshefte*, **1972** (9), 538–545.

LEWANDOWSKI, M. 1993. Paleomagnetism of the Paleozoic rocks of the Holy Cross Mts (Central Poland) and the Origin of the Variscan Orogeny. *Publications of the Institute of Geophysics, Polish Academy of Sciences*, **A-23** (265), 1–85.

LIN BAOYU, TCHI YONGYI, JIN CHUNTAI, LI YAOXI and YAN YOUYIN 1988. *Monography of Palaeozoic corals. Vol. II: Tabulatomorphic corals*. Geological publishing House, Beijing, 467 pp. [In Chinese, with English summary].

LITTLEWOOD, D. T. J. and DONOVAN, S. K. 2003. Fossil parasites: a case of identity. *Geology Today*, **19**, 136–142.

ŁOBANOWSKI, H. 1971. The Lower Devonian in the western part of the Klonów belt (Holy Cross Mts.). Part 1. *Acta Geologica Polonica*, **21**, 629–687.

LUKIN, V. Y. 1998. Devonian tabulates of Polar Urals (Lek-Elec river). *Syktyvkarski Paleontologicheskii Sbornik*, **3** (*Trudy Instituta geologii Komi Nauchnogo Centra UrO Rossijskoj Akademii Nauk*, **99**), 20–34. [In Russian].

LÜTTE, B.-P. 1993. Zur stratigraphischen Verteilung tabulater Korallen im Mittel-Devon der Sötenicher Mulde (Rheinisches Schefergebirge, Nord-Eifel). *Geologica et Palaeontologica*, **27**, 55–71.

MALEC, J. and RACKI, G. 1992. Givetian and Frasnian ostracod associations from the Holy Cross Mountains. *Acta Palaeontologica Polonica*, **37**, 359–384.

MATÉ, J. L. 2003. Ecological, genetic, and morphological differences among three *Pavona* (Cnidaria: Anthozoa) species from the Pacific coast of Panama. *Marine Biology*, **142**, 427–440.

MAY, A. 1993a. Korallen aus dem Höheren Eifelium und Unteren Givetium (Devon) des Nordwestlichen Sauerlandes (Rheinisches Schiefergebirge). Teil 1: Tabulate Korallen. *Palaeontographica, Abteilung A*, **227**, 87–224.

—— 1993b. *Thamnopora* und verwandte ästige tabulate Korallen aus dem Emsium bis Unter-Eifelium von Asturien (Devon; Nord-Spanien). *Geologica et Palaeontologica*, **27**, 73–101.

—— 1995. *Thamnopora* (Anthozoa; Tabulata) aus dem Givetium bis Frasnium von Asturien (Devon; Nord-Spanien). *Münster Forschungen zur Geologie und Paläontologie*, **77**, 479–491.

—— 1997. Statistics on *Thamnopora* (Tabulata, Devonian). *Boletín de la Real Sociedad Española de Historia Natural*, **91**, 217–230.

—— 1998. Statistische Untersuchungen an der tabulaten Koralle *Thamnopora* (Anthozoa; Devon). *Geologica et Palaeontologica*, **32**, 141–159.

—— 2006. Micheliniidae and Cleistoporidae (Anthozoa, Tabulata) from the Devonian of Spain. *Bulletin of Geosciences*, **81**, 163–172.

—— 2007. Reply to the Comments of Yves Plusquellec and Esperanza Fernández-Martínez on the paper by A. May "Micheliniidae and Cleistoporidae (Anthozoa, Tabulata) from the Devonian of Spain". *Bulletin of Geosciences*, **82**, 90–94.

MAYR, E. 1942. *Systematics and the origin of species from the viewpoint of a zoologist*. Columbia University Press, New York.

MCCOY, F. 1850. Description of three new Devonian zoophytes. *Annals and Magazine of Natural History*, **2**, 377–378.

MICHALSKI, A. 1888a. Sprawozdanie przedwstępne z badań dokonanych w południowéj części gubernii radomskiéj. *Pamiętnik Fizyograficzny*, **8** (2), 37–45.

—— 1888b. Compte-rendu préliminaire sur les recherches géologiques faites dans la partie méridionale du gouvernement de Radom. *Izviestia Geologicheskoi Kommisii*, **7**, 177–191. [In Russian, French summary].

MICHELIN, H. 1840. *Iconographie zoophytologique, descriptions par localités et terrains des polypiers fossils de France et pays environnants*. Librairie Bertrand, Paris, 348 pp.

MILLER, K. J. and BENZIE, J. A. H. 1997. No clear genetic distinction between morphological species within the coral genus *Platygyra*. *Bulletin of Marine Science*, **61**, 907–917.

MILNE-EDWARDS, H. and HAIME, J. 1850. *A monograph of British fossil corals*. 5 volumes. Palaeontographical Society Monographs, London.

—— —— 1851. Monographie des polypiers fossiles des terrains paléozoïques. *Archives du Muséum d'Histoire Naturelle de Paris*, **5**, 1–502, pls 1–20.

MIRONOVA, N. V. 1965. On the problem of genetic variability in some favositid genera. 79–86. *In* SOKOLOV, B. S. and DUBATOLOV, V. N. (eds). *Tabulatomorph corals from the Devonian and Carboniferous of the Soviet Union*. Trudy 1. Vsesoyuznego simpoziuma po izucheniyu iskopaemych korallov, **2**. Moscow, Nedra, 138 pp. [In Russian]

—— 1970. New genera of Alveolitidae (Tabulata). *Trudy SNIIGGIMS*, **110**, 126–130. [In Russian]

—— 1974. *Early Devonian Tabulates of the Upper Altai and Salair*. Ministerstvo Gieologii SSSR, SNNIIGIMS, Novosibirsk, 174 pp. [In Russian]

MISTIAEN, B. 1984. Comments on the caunopore tubes: stratigraphic distribution and microstructure. *Palaeontographica Americana*, **54**, 501–508.

—— 1988. Tabulés Auloporida du Givétien et du Frasnien de Ferques (Boulonnais, France). *In* BRICE, D. (ed.). *Le Dévonien de Ferques. Bas-Boulonnais (N. France). Biostratigraphie du Paléozoïque*, **7**, 197–230.

—— 1989. Importance de la symétrie d'ordre douze chez *Tabulata*. *Comptes Rendus de l'Academie des Sciences, Paris, Series II*, **308**, 451–456.

—— 2002. Stromatopores et coraux tabulés du Membre de Pâtures, Formation de Beaulieu (Frasnien de Ferques, Boulonnais). *Annales de la Société Géologique du Nord, 2ème Série*, **9**, 85–90.

—— BECKER, T., BRICE, D., DÉGARDIN, J.-M., DERYCKE, C., LOONES, C. and ROHART, J.-C. 2002. Données nouvelles sur la partie supérieure de la Formation de Beaulieu (Frasnien de Ferques, Boulonnais). *Annales de la Société Géologique du Nord, 2ème Série*, **9**, 75–84.

MIZERSKI, W. 1995. Geotectonic evolution of the Holy Cross Mts in central Europe. *Biuletyn Państwowego Instytutu Geologicznego*, **327**, 1–47.

—— 2004. Holy Cross Mountains in the Caledonian, Variscan and Alpine cycles – major problems, open questions. *Przegląd Geologiczny*, **52**, 774–779.

MÕTUS, M.-A. 2001. Environment related morphological variation in Early Silurian tabulate corals from the Baltic Sea. *Bulletin of the Tohoku University Museum*, **1**, 62–69.

—— 2006. Intraspecific variation in Wenlock tabulate corals from Saaremaa (Estonia) and its taxonomic implication. *Proceedings of the Estonian Acadademy Sciences, Geology*, **55**, 24–42.

NARKIEWICZ, M., RACKI, G. and WRZOŁEK, T. 1990. Lithostratigraphy of the Devonian stromatoporoid-coral carbonate sequence in the Holy Cross Mountains. *Kwartalnik Geologiczny*, **34**, 433–456.

NAWROCKI, J. and POPRAWA, P. 2006. Development of Trans-European Suture Zone in Poland: from Ediacaran rifting to Early Palaeozoic accretion. *Geological Quarterly*, **50**, 59–76.

NICHOLSON, H. A. 1879. *On the Structure and Affinities of the "Tabulate Corals" of the Paleozoic Period with Critical Descriptions of Illustrative Species*. William Blackwood and Sons, Edinburgh and London. x + 342 pp., 15 pls.

NIKO, S. 2001. Late Silurian Auloporids (Coelenterata: Tabulata) from the Hitoegane Formation, Gifu Prefecture. *Bulletin of the National Science Museum, Series C*, **27** (3–4), 63–71.

—— 2003. Devonian Coenitid Tabulate Corals from the Fukuji Formation, Gifu Prefecture. *Bulletin of the National Science Museum, Series C*, **29** (1), 19–24.

—— and ADACHI, T. 1999. *Gokaselites*, a New Genus of Silurian Tabulate Coral from the Gionyama Formation, Miyazaki Prefecture. *Bulletin of the National Science Museum, Series C*, **25** (1–2), 45–49.

NOBLE, J. P. A. and YOUNG, G. A. 1984. The Llandovery-Wenlock heliolitid corals from New Brunswick, Canada. *Journal of Paleontology*, **58**, 867–884.

NOTHDURFT, L. D. and WEBB, G. E. 2007. "Shingle" microstructure in scleractinian corals: a possible analogue for lamellar and microlamellar microstructure in Palaeozoic tabulate corals. *Österreichischen Akademie der Wissenschaften, Schriftenreihe der Erdwissenschaftlichen Kommissionen*, **17**, 85–99.

NOWIŃSKI, A. 1970. *Syringella* – a new genus of the family Syringoporidae (Tabulata) from the Devonian of Poland. *Acta Palaeontologica Polonica*, **15**, 539–544.

—— 1976. Tabulata and Chaetetida from the Devonian and Carboniferous of southern Poland. *Palaeontologia Polonica*, **35**, 1–125.

—— 1991. Late Carboniferous to Early Permian Tabulata from Spitsbergen. *Palaeontologia Polonica*, **51**, 1–74.

—— 1992. Tabulate corals from the Givetian and Frasnian of the Holy Cross Mountains and Silesian Upland. *Acta Palaeontologica Polonica*, **37**, 183–216.

—— 2003. Podgromada Tabulata Milne Edwards et Haime, 1850. 124–166. *In* MALINOWSKA, L. (Ed.), *Budowa Geologiczna Polski, vol. 3, part 1b. Atlas skamieniałości przewodnich i charakterystycznych. Dewon*. Państwowy Instytut Geologiczny, Warszawa, 897 pp., 403 pls.

—— and PREJBISZ, A. 1986. Devonian tabulate corals from western Pomerania, Poland. *Acta Palaeontologica Polonica*, **31**, 237–251.

OEKENTORP, K. 1969. Kommensalismus bei Favositiden. *Münster Forschungen zur Geologie und Paläontologie*, **12**, 165–217.

—— 1972. Sekundärstrukturen bei paläozoischen Madreporaria. *Münster Forschungen zur Geologie und Paläontologie*, **24**, 35–108.

—— 1980. Aragonit und Diagenese bei jungpaläozoischen Korallen. *In* OEKENTORP, K. (ed.). *Festschrift Professor Dr. Alexander von Schouppé. Münster Forschungen zur Geologie und Paläontologie*, **52**, 119–239.

—— 1985. Spicules in favositid Tabulata – remarks to J. Kaźmierczak's interpretation. *Fossil Cnidaria*, **14**, 34–35.

—— 1989. Diagenesis in corals: syntaxial cements as evidence for post-mortem skeletal thickenings in Palaeozoic corals.

Memoir of the Association of Australasian Palaeontologists, **8**, 169–177.

—— 2001. Review on diagenetic microstructures in fossil corals – a controversial discussion. *Bulletin of the Tohoku University Museum*, **1**, 193–209.

—— 2007. The microstructure concept – coral research in the conflict of controversial opinions. *Bulletin of Geosciences*, **82**, 95–97.

—— and BRÜHL, D. 1999. Tabulaten-Fauna im Grenzbereich Unter-/Mittel-Devon der Eifeler Richtschnitte (S-Eifel/ Rheinisches Schiefergebirge). *Senckenbergiana lethaea*, **79**, 63–87.

—— and STEL, J. 1985. *Favosites* – a true coral. Remarks to P. Copper's discoveries of fossilized polyps. *Fossil Cnidaria*, **14**, 28–29.

OLEMPSKA, E. 1979. Middle to Upper Devonian Ostracoda from the Southern Holy Cross Mountains, Poland. *Palaeontologia Polonica*, **40**, 57–162.

OLIVER, W. A. 1966. Description of dimorphism in *Striatopora flexuosa* Hall. *Palaeontology*, **9**, 448–454.

—— 1979. Sponges, they are not. *Paleobiology*, **5**, 188–190.

—— 1986. Favositids are corals – Further remarks. *Fossil Cnidaria*, **15**, 19–21.

—— 1989. Intraspecific variation in pre-Carboniferous rugose corals: a subjective review. *Memoir of the Association of Australasian Palaeontologists*, **8**, 1–6.

OSPANOVA, N. K. and LELESHUS, V. L. 1988. On the first find of *Adetopora* (Tabulata) in the Lower Silurian of Tajikistan. *Doklady Akademii Nauk Tadzhikskoi SSR*, **31** (2), 134–136.

PAJCHLOWA, M. 1957. Dewon w profilu Grzegorzowice-Skały. *Biuletyn Instytutu Geologicznego*, **122**, 145–254.

PANDOLFI, J. M. 1989. Developmental sequences in colonial corals: an overview. *Memoir of the Association of Australasian Palaeontologists*, **8**, 69–81.

PERRIN, C. and CUIF, J.-P. 2001. Ultrastructural controls on diagenetic patterns of scleractinian skeletons: evidence at the scale of colony lifetime. *Bulletin of the Tohoku University Museum*, **1**, 210–218.

—— and SMITH, D. C. 2007. Decay of skeletal organic matrices and early diagenesis in coral skeletons. *Comptes Rendus Palevol*, **6**, 253–260.

PHILLIPS, J. 1841. *Figures and descriptions of the Palaeozoic fossils of Cornwall, Devon and West Somerset*. Geological Survey of Great Britain and Ireland, London, 231 pp.

PISARZOWSKA, A., SOBSTEL, M. and RACKI, G. 2006. Conodont–based event stratigraphy of the Early–Middle Frasnian transition on the South Polish carbonate shelf. *Acta Paleontologica Polonica*, **51**, 609–646.

PLUSQUELLEC, Y. 1968. Commensaux des Tabulés et Stromatoporoïdes du Dévonien armoricain. *Annales de la Societe Géologique du Nord*, **88**, 47–56.

—— 2007. Histoire naturelle des pleurodictyiformes (Cnidaria, Tabulata, Dévonien) du Massif Armoricain et des regions maghrebo-européennes principalement. *Mémoires de la Société Géologique et Mineralogique de Bretagne*, **32**, 1–123.

—— and FERNÁNDEZ-MARTÍNEZ, E. 2007. Comments on the paper by A. May "Michelinidae and Cleistoporidae

(Anthozoa, Tabulata) from the Devonian of Spain". *Bulletin of Geosciences*, **82**, 85–89.

—— and SANDO, W. J. 1987. The microstructure of *Michelinia meekana* Girty, 1910. *Journal of Paleontology*, **61**, 10–13.

—— and TCHUDINOVA, I. I. 1977. The microstructure of *Parastriatopora* Sokolov, 1949 (Siluro-Devonian Tabulata). *Annales de la Société Géologique du Nord*, **97**, 127–130.

—— TOURNEUR, F. and GROISARD, E. 1997. Revision of *Alveolites fougti* Milne-Edwards & Haime, 1851, type species of *Planalveolites* Lang & Smith, 1939 (Tabulata, Silurian of Gotland). *Boletín de la Real Sociedad Española de Historia Natural*, **91**, 205–215.

—— FERNÁNDEZ-MARTÍNEZ, E., MISTIAEN, B. and TOURNEUR, F. 2004. Révision de *Crenulipora difformis* Le Maître, 1956 (Tabulata, Dévonien du Nord Gondwana): morphologie, structure et microstructure. *Revue de Paléobiologie*, **23**, 181–208.

—— TOURNEUR, F. and LE HÉRISSÉ, A. 2007. Structure and microstructure of *Pachypora lamellicornis* Lindström, 1873, a tabulate coral from the Silurian of Gotland, Sweden. *In* HUBMANN, B. and PILLER, W. E. (eds). *Fossil Corals and Sponges: Proceedings of the 9th International Symposium on Fossil Cnidaria and Porifera, Graz 2003 – Österreichische Akademie der Wissenschaften. Schriftenreihe der Erdwissenschaftlichen Komisionen*, **17**, 67–83.

POTTS, D. C., BUDD, A. F. and GARTHWAITE, R. L. 1993. Soft tissue vs. skeletal approaches to species recognition and phylogeny reconstruction in corals. *Courier Forschungsinstitut Senckenberg*, **164**, 221–231.

POWELL, J. H. and SCRUTTON, C. T. 1978. Variation in the Silutian tabulate coral *Paleofavosites asper*, and the status of *Mesofavosites*. *Palaeontology*, **21**, 307–319.

PUSCH, G. G. 1837. *Polens Paläontologie*. E. Schweizerbart's Verlaghandlung, Stuttgart, 128 pp.

RACKI, G. 1992*a*. Evolution of the bank to reef complex in the Devonian of the Holy Cross Mountains. *Acta Palaeontologica Polonica*, **37**, 87–182.

—— 1992*b*. Brachiopod assemblages in the Devonian Kowala Formation of the Holy Cross Mountains. *Acta Palaeontologica Polonica*, **37**, 297–357.

—— 1997. Devonian eustatic fluctuations in Poland. *Courier Forschungsinstitut Senckenberg*, **199**, 1–12.

RADUGIN, K. V. 1938. Coelenterata of the Middle Devonian of Lebediansk area. *Izviestia Tomskogo Industrialnogo Instituta*, **56**, 49–109. [In Russian]

REED, C. F. R. 1908. The Devonian faunas of the northern Shan states. *Memoirs of the Geological Survey of India – Palaeontologia Indica – New Series*, **2**, 1–183, 20 pls.

REITNER, J. 1989. Some remarks to J. Kaźmierczak's "spicules" in *Quepora agglomeratiformis* (Whitfield). *Fossil Cnidaria*, **18**, 22–24.

RIGBY, J. K., PISERA, A., WRZOŁEK, T. and RACKI, G. 2001. Upper Devonian sponges from the Holy Cross Mountains, central Poland. *Palaeontology*, **44**, 447–488.

ROEMER, C. F. 1844. *Das Rhenischen Ubergangsgebirge*. Hahn'schen Hofbuchhandlung, Hannover, 96 pp.

—— 1883. Lethaea Geognostica oder Beschreibung und Abbildung der für die Gebirgs-Formation bezeichnendsten Vereinerungen. 1. *Lethaea Palaeozoica*, **1**, 1–668.

ROEMER, F. 1866. Geognostische Beobachtungen im Polnischen Mittelgebirge. *Zeitschrift der Deutschen Geologischen Gesselschaft*, **19**, 667–690.

RÓŻKOWSKA, M. 1953. Pachyphyllinae et *Philipsastraea* du Frasnien de la Pologne. *Palaeontologia Polonica*, **5**, 1–89.

—— 1979. Contribution to the Frasnian tetracorals from Poland. *Palaeontologia Polonica*, **40**, 1–56.

SARDESON, W. 1896. Ueber die Beziehungen der Fossilen Tabulaten zu den Alcyonarien. *Neues Jahrbuch für Mineralogie, Beilband*, **10**, 249–362.

SARNECKA, E. 1987. Tabulata i Chaetetida w wybranych profilach eiflu Gór Świętokrzyskich. *Biuletyn Instytutu Geologicznego*, **354**, 125–144.

—— 1997. Tabulata from the Lower and Middle Devonian of the Holy Cross Mts. *Geological Quarterly*, **41**, 151–168.

SCHINDEWOLF, O. H. 1959. Würmer und Korallen als Synöken. Zur Kenntnis der Systeme *Aspidosyphon*/Heteropsammia und *Hicetes*/Pleurodictyum. *Abhandlungen der Akademie der Wissenschaften und der Literatur, Matematisch-Naturwissenschaftlische Klasse*, **6**, 263–327. [for 1958].

SCHLÜTER, C. 1889. Anthozoen des rheinischen Mittel-Devon. *Abhandlungen zur Geologischen Specialkarte von Preussen und den Thüringischen Staaten*, **8** (4), x + 1–207.

SCHMIDT, O. 1868. *Die Spongien der Küste von Algier. Mit Nachträgen zu den Spongien des Adriatischen Meeres (drittes Supplement)*. Wilhelm Engelmann, Leipzig, iv + 44 pp.

SCRUTTON, C. T. 1981. The measurement of corallite size in corals. *Journal of Paleontology*, **55**, 687–688.

—— 1987. A review of favositid affinities. *Palaeontology*, **30**, 485–492.

—— 1989. Intracolonial and intraspecific variation in tabulate corals. *Memoir of the Association of Australasian Palaeontologists*, **8**, 33–43.

—— 1990. Ontogeny and astogeny in *Aulopora* and its significance, illustrated by a new, non-encrusting species from the Devonian of southwest England. *Lethaia*, **23**, 61–65.

—— 1997. The Palaeozoic corals, I: origins and relationships. *Proceedings of the Yorkshire Geological Society*, **51**, 177–298.

—— and POWELL, J. H. 1980. Periodic development of dimetrism in some favositid corals. *Acta Palaeontologica Polonica*, **25**, 477–491.

SIMPSON, G. G., ROE, A. and LEWONTIN, R. C. 1960. *Quantitative Zoology*. Harcourt, Brace & World, Inc., New York, 440 pp.

SMITH, S. 1933. Sur des espèces d'*Alveolites* de l'Eifelien inférieur du Nord de la France et de la Belgique. *Annales de la Société Géologique du Nord*, **58**, 134–145.

SOBOLEV, D. 1904. Devonian beds of the section Grzegorzovice-Skaly-Vlohy. *Izdanie Varshavskago Politechnicheskago Instituta*, (volume, issue unknown), 1–107, 9 pls. [In Russian]

SOKOLOV, B. S. 1947. New syringoporids of Taymyr. *Biulletin Moskovskogo Obshchestva Ispytatelei Prirody, (Geologiia)*, **22** (6), 19–28. [In Russian].

—— 1948. Commensalism among the favositids. *Izviestia Akademii Nauk SSSR, Seria Biologicheskaya*, **1**, 101–110. [In Russian].

—— 1950. Silurian corals of the western part of the Siberian Platform. *Vaprosi Paleontologii*, **1**, 211–242. [In Russian].

—— 1952. Tabulata of the Palaeozoic of the European part of the USSR. Part IV. The Devonian of the Russian Platform and western Urals. *Trudy VNIGRI*, **62**, 1–208. [In Russian].

—— 1955. Tabulata of the Palaeozoic of the European part of the USSR. Introduction: general problems of systematics and history of tabulates. *Trudy VNIGRI*, **85**, 1–527. [In Russian].

—— 1962a. Tabulata. 122–285. *In* ORLOV, Y. A. (ed.). *Osnovy Paleontologii. Gubki, Arkheotsyaty, Kischechnopolyustnye, Chervi*. Nauka, Moscow, 463 pp. [In Russian].

—— 1962b. A widespread commensal associate of Devonian favositids. *Paleontologicheskii Zhurnal*, **1962** (2), 45–48. [In Russian].

SORAUF, J. E. 1997. Geochemical signature of incremental growth and diagenesis of skeletal structure in *Tabulophyllum traversensis* (Winchell, 1866). *Boletín de la Real Sociedad Española de Historia Natural*, **91**, 77–86.

—— and MACKEY, S. D. 1989. Variation and biometrics in rugose corals. *Memoir of the Association of Australasian Palaeontologists*, **8**, 23–31.

—— and STEIN, W. E. Jr 1993. Biological fabric and the study of growth in the Devonian tabulate coral genera *Lecfedites* and *Favosites*. *Courier Forschungsinstitut Senckenberg*, **164**, 159–168.

—— and WEBB, G. E. 2003. The origin and significance of zigzag microstructure in Late Palaeozoic *Lophopyllidium* (Anthozoa, Rugosa). *Journal of Paleontology*, **77**, 16–30.

SPRIESTERBACH, J. 1935. Beitrag zur Kenntnis der Fauna der rheinischen Devon. *Jahrbuch der Preußichen Geologischen Landesanstalt*, **55**, 475–525. [Dated 1934].

STADELMAIER, M., NOSE, M., MAY, A., SALERNO, C., SCHRÖDER, S. and LEINFELDER, R. R. 2005. Ästige tabulate Korallen-Gemeinschaften aus dem Mitteldevon der Sötenicher Mulde (Eifel): Faunenzusammensetzung und fazies Umfeld. *Zittieliana*, **B25**, 5–38.

STASIŃSKA, A. 1953. Genus Alveolites Lamarck from Devonian of the Holy Cross Mountains. *Acta Geologica Polonica*, **3**, 211–237.

—— 1954. Koralowce Tabulata z dewonu Grzegorzowic (badania wstępne). *Acta Geologica Polonica*, **4**, 277–290.

—— 1958. Tabulata, Heliolitida et Chaetetida du Devonien moyen des Monts de Sainte-Croix. *Acta Palaeontologica Polonica*, **3**, 161–241.

—— 1969a. Structure and ontogeny of *Kozlowskiocysta polonica* (Stasińska, 1958). *Acta Palaeontologica Polonica*, **14**, 553–560.

—— 1969b. Koralowce dewońskie Tabulata z otworu Miastko-1 w północno-zachodniej Polsce. *Acta Geologica Polonica*, **19**, 265–280.

—— 1974. On some Devonian Auloporida (Tabulata) from Poland. *Acta Palaeontologica Polonica*, **19**, 265–280.

—— and NOWIŃSKI, A. 1976. Tabulata from the Givetian of the south-eastern Poland. *Acta Palaeontologica Polonica*, **21**, 293–309.

—— —— 1978. Frasnian Tabulata of the south-eastern Poland. *Acta Palaeontologica Polonica*, **23**, 199–218.

STEININGER, J. 1831. *Bemerkungen über die Versteinerungen, welche in dem Uebergangs-Kalkgebirge der Eifel gefunden werden*. Gymnasium zu Trier, Trier, 44 pp.

—— 1849. Die Versteinerungen des Uebergangsgebirges der Eifel. *In* LOERS, V. (ed.). *Jahresbericht über den Schul-Cursus 1848/49 an dem Gymnasium zu Trier*. Gymnasium zu Trier, Trier, 34 pp.

STEL, J. H. 1976. The Palaeozoic hard substrate trace fossils *Helicosalpinx*, *Chaetosalpinx* and *Torquaysalpinx*. *Neues Jahrbuch für Geologie und Paläontologie, Monatshefte*, **1976**, 726–744.

—— 1978. *Studies on the Palaeobiology of Favositids*. Stabo/all-round, Groeningen, IV + 246 pp.

STOLARSKI, J. 1990. On Cretaceous *Stephanocyathus* (Scleractinia) from the Tatra Mts. *Acta Palaeontologica Polonica*, **35**, 31–39.

STUMM, E. C. 1960. The type species of the Paleozoic tabulate coral genera *Cladopora* and *Coenites*. *Contributions from the Museum of Paleontology*, **15** (7), 133–138.

—— 1965. Silurian and Devonian Corals of the Falls of the Ohio. *The Geological Society of America, Memoir*, **93**, 1–184.

—— 1967. *Planalveolitella*, a new genus of Devonian tabulate corals, with a redescription of *Planalveolites* foughti (Edwards and Haime). *Contributions from the Museum of Paleontology*, **21** (2), 67–72.

STUPNICKA, E. 1992. The significance of the Variscan orogeny in the Świętokrzyskie Mountains (Mid Polish Uplands). *Geologische Rundschau*, **81**, 561–570.

SUGIYAMA, T. 1940. Stratigraphical and palaeontological notes of the Gotlandian deposits of the Kitakami Mountainland. *Scientific Reports of the Tohoku University, Series 2*, **21**, 81–146.

SUTHERLAND, P. K. 1989. Intraspecific variation in the rugose coral *Stelechophyllum? mclareni* from the Lower Carboniferous (Visean) of northeastern British Columbia. *Memoir of the Association of Australasian Palaeontologists*, **8**, 13–22.

SUTTON, I. D. 1966. The value of corallite size in the specific determination of the tabulate corals *Favosites* and *Palaeofavosites*. *Mercian Geologist*, **1**, 255–263.

SZULCZEWSKI, M. 1995. Depositional evolution of the Holy Cross Mts. (Poland) in the Devonian and Carboniferous – a review. *Geological Quarterly*, **39**, 471–488.

—— BELKA, Z. and SKOMPSKI, S. 1996. The drowning of a carbonate platform: an example from the Devonian-Carboniferous of the southwestern Holy Cross Mountains, Poland. *Sedimentary Geology*, **106**, 21–49.

TAPANILA, L. 2002. A New Endosymbiont in Late Ordovician Tabulate Corals from Anticosti Island, Eastern Canada. *Ichnos*, **9**, 109–116.

—— 2004. The earliest *Helicosalpinx* from Canada and the global expansion of commensalism in Late Ordovician sarcinulid corals (Tabulata). *Palaeogeography Palaeoclimatology, Palaeoecology*, **215**, 99–110.

—— 2005. Palaeoecology and diversity of endosymbionts in Palaeozoic marine invertebrates: trace fossil evidence. *Lethaia*, **38**, 89–99.

TARLO, L. B. H. 1964. Psammosteiformes (Agnatha). 1. General part. *Palaeontologia Polonica*, **13**, 1–135.

—— 1965. Psammosteiformes (Agnatha). 2. Systematic part. *Palaeontologia Polonica*, **15**, 1–164.

TCHERNYCHEV, V. V. 1937. Silurian and Devonian Tabulata of Mongolia and Tuva. *Trudy Mongolskoi Komissi*, **30**, 1–34.

—— 1938. Tabulata of the Vaygach Island. *Transactions of the Arctic Institute*, **101**, 109–155. [In Russian].

—— 1941. Tabulata of the Main Devonian Field. Fauna of the Main Devonian Field. *Izdanie AN SSSR*, **1**, 113–131. [In Russian].

TCHI, Y.-Y. 1980. Tabulata. 153–188. *In* Paleontological atlas of northeast China. 1. Paleozoic Volume. Geological Publishing House, Beijing. [In Chinese].

—— 1987. A study of Devonian Tabulata, Heliolitida and Chaetetida from Luqu-Tewo District., West Quiling Mountains. 280–290. *In* Late Silurian–Devonian Strata and Fossils from Luqu-Tewo District, West Quiling Mountains, China. Xi'an Institute of Geology and Mineral Resources, Nanjing Institute of Geology and Palaeontology, Academia Sinica. Nanjing University Press, Nanjing, 741 pp, [in Chinese]

TCHUDINOVA, I. I. 1959. Devonian thamnoporids of the southern Siberia. *Trudy Paleontologitscheskogo Instituta*, **73**, 1–144. [In Russian].

—— 1964. Tabulata of the Lower and Middle Devonian of the Kuznetsk Basin. *Trudy Paleontologitscheskogo Instituta*, **101**, 3–79. [In Russian].

—— 1986. *Composition, System and Filogeny of the Fossil Corals: Order Syringoporida*. Nauka, Moscow, 204 pp. [In Russian].

TODD, P. A. 2008. Morphological plasticity in scleractinian corals. *Biological Reviews*, **83**, 315–337.

TONG-DZUY, T. 1967. Les coelenteres du Dévonien au Viet Nam. Partie I. Les coraux tabulatomorphes du Dévonien au Nord Viet Nam. *Acta Scientarum Vietnamicarum, Sectio Scientarum Geologicarum et Geographicarum*, **3**, 1–304.

—— NGUYEN, D. C. and KHROMYKH, V. G. 1988. Devonian stratigraphy and coelenterata of Vietnam. Vol. 2: Coelenterata. *In* DUBATOLOV, V. N. and TESAKOV, YU. I. (eds.). *Devonian stratigraphy and coelenterata of Vietnam*. Nauka, Novosibirsk, 248 pp.

TOURNEUR, F. 1985. Contribution a l'étude des Tabulés du Dévonien moyen de la Belgique. Unpublished PhD Thesis, Université Catholique de Louvain, Louvain-la-Neuve, 568 pp., 57 pls.

—— 1986. Microstructure des genres *Thamnopora* Steininger, 1831 et *Lecomtopora* nov. gen., Tabulés branchus du Dévonien moyen. *Comptes Rendus de l'Academie des Sciences, Séries II.*, **303**, 1255–1258.

TSCHERNYCHEV, T. 1887. Die fauna des Mitteleren und Obern Devon am West-Abhange des Urals. *Memoires du Comité Géologique*, **3** (3), 1–208, 14 pls.

TSUKADA, K. 2005. Tabulate corals from the Devonian Fukuji Formation, Hida Gaien belt, central Japan, Part 1. *Bulletin of the Nagoya University Museum*, **21**, 57–125.

TSYGANKO, V. S. and LUKIN, V. YU. 2005. Tabulata and Rugosa of Uhtinsk Anticline (Southern Timan). *Syktyvkarski Paleontologicheskii Sbornik*, **6** (*Trudy Instituta geologii Komi*

Nauchnogo Centra UrO Rossijskoj Akademii Nauk, **117**), 14–48. [In Russian].

VERON, J. E. N. 1993. *Corals of Australia and the Indo-Pacific.* University of Hawaii Press, Honolulu, 644 pp.

——2007. Corals: pointing to a different evolution. *Österreichischen Akademie der Wissenschaften, Schriftenreiche der Erdwissenschaftlichen Kommisionen,* **17**, 507–515.

WATKINS, R. 2000. Corallite size and spacing as an aspect of niche-partitioning in tabulate corals of Silurian reefs, Racine Formation, North America. *Lethaia,* **33**, 55–63.

WEDEKIND, R. 1937. *Einführung in die Grundlagen der historischen Geologie, II. Band. Mikrobiostratigraphie, Die Korallen- und Foraminiferenzeit.* Ferdinand Enke, Stuttgart, 136 pp, 16 pls.

WOOD, R., COPPER, P. and REITNER, J. 1990. "Spicules" in halysitids: a reply. *Lethaia,* **23**, 113–114.

WRZOŁEK, T. 1992. Rugose corals from the Devonian Kowala Formation of the Holy Cross Mountains. *Acta Palaeontologica Polonica,* **37**, 217–254.

YANET, F. E. 1972. Subclass Tabulata. 48–97. *In* KODALEVITCH, A. N. (ed.). *Coelenterata and Brachiopoda from the Givetian deposits of western slopes of Ural.* Ministerstvo Gieologii SSSR, Moscow, 263 pp. [In Russian].

YOUNG, G. A. and ELIAS, R. J. 1993. Biometry and intraspecific variation in favositid and heliolitid corals. *Courier Forschungsinstitut Senckenberg,* **164**, 283–291.

—— —— 1995. Latest Ordovician to earliest Silurian colonial corals of the east-central United States. *Bulletins of American Paleontology,* **108**, 1–148.

—— —— 1997. Patterns of variation in Late Ordovician and Early Silurian tabulate corals. *Boletín de la Real Sociedad Española de Historia Natural,* **91**, 193–204.

—— and NOBLE, J. P. A. 1989. Variation and growth of a syringoporid symbiont species in stromatoporoid from the Silurian of eastern Canada. *Memoir of the Association of Australasian Palaeontologists,* **8**, 91–98.

ZAPALSKI, M. K. 2003. Auloporida (Tabulata) from the Devonian of the Holy Cross Mts, Poland. Unpublished MSc Thesis. Warsaw University, Warszawa, 52 pp, 4 pls.

——2004. Parasitism on favositids (Tabulata). *The Palaeontological Association Newsletter,* **57**, 194.

——2005a. A new species of Tabulata from the Emsian of the Holy Cross Mts., Poland. *Neues Jahrbuch für Geologie und Paläontologie – Monatshefte,* **2005**, 248–256.

——2005b. Paleoecology of Auloporida: an example from the Devonian of the Holy Cross Mts. Poland. *Geobios,* **38**, 677–683.

——2007a. Parasitism versus commensalism – the case of tabulate endobionts. *Palaeontology,* **50**, 1375–1380.

——2007b. Growth pattern of *Alveolites compressus* from the Upper Devonian of the Holy Cross Mts., Poland. 105. *In* KOSSOVAYA, O., SOMERVILLE, I. and EVDOKIM-

OVA, I. (ed.). *X International congress on Fossil Cnidaria and Porifera. Abstracts.* VSEGEI, St. Petersburg, 110 pp.

——2007c. Taxonomical value of selected biometrical characters: example of *Alveolites* (Tabulata) from the Frasnian of the Holy Cross Mts. (Poland). *The Palaeontological Association Newsletter,* **66**, 102.

—— 2009. Parasites in Emsian–Eifelian Favosites (Anthozoa, Tabulata) from the Holy Cross Mountains (Poland): changes of distribution within colony. *In* KÖNIGSHOF, P. (ed.). *Palaeozoic Reefs and Bioaccumulations: Devonian Change. Case Studies in Palaeogeography and Palaeoecology.* Geological Society, London, Special Publications, **314**, 125–129.

——2011. Is absence of proof a proof of absence? Comments on commensalism. *Palaeogeography, Palaeoclimatology, Palaeoecology,* **302**, 484–488.

—— and BERKOWSKI, B. in press. The oldest species of *Yavorskia* (Tabulata) from the late Famennian of the Holy Cross Mountains (Poland). *Acta Geologica Polonica,* (in press).

—— and HUBERT, B. L. M. 2011. First fossil record of parasitism in Devonian calcareous sponges (stromatoporoids). *Parasitology,* **138**, 132–138.

—— and NOWIŃSKI, A. 2005. *Maksymilianites,* a new name for Syringella Nowiński, 1970 (Anthozoa, Tabulata) preoccupied by Syringella Schmidt, 1868 (Porifera). *Paläontologische Zeitschrift,* **79**, 507–508.

—— —— 2011. A new Silurian Avicenia (Tabulata): taxonomy, growth pattern and colony integration. *Geodiversitas,* **33**, 541–551.

——HUBERT, B. L. M. and MISTIAEN, B. 2007a. Estimation of palaeoenvironmental changes: can analysis of distribution of tabulae in tabulates be a tool? *In* ÁLVARO, J. J., ARETZ, M., BOULVAIN, F., MUNNECKE, A., VACHARD, D. and VENNIN, E. le. (eds). *Palaeozoic Reefs and Bioaccumulations: Climatic and Evolutionary Controls.* Geological Society, London, Special Publications, **275**, 275–281.

—— —— NICOLLIN, J.-P., MISTIAEN, B. and BRICE, D. 2007b. The palaeobiodiversity of stromatoporoids, tabulates and brachiopods in the Devonian of the Ardennes – changes through time. *Bulletin de la Société Géologique de France,* **178**, 383–390.

—— PINTE, E. and MISTIAEN, B. 2008. Late Famennian *?Chaetosalpinx* in *Yavorskia* (Tabulata): the youngest record of tabulate endobionts. *Acta Geologica Polonica,* **58**, 321–324.

—— TRAMMER, J. and MISTIAEN, B. 2012. Unusual growth pattern in the Frasnian alveolitids (Tabulata) from the Holy Cross Mts. (Poland). *Palaeontology,* doi: 10.1111/j.1475-4983.2012.01141.x.

ZLATARSKI, V. N. 2007. The scleractinian species – a holistic approach. *Österreichischen Akademie der Wissenschaften, Schriftenreiche der Erdwissenschaftlichen Kommisionen,* **17**, 523–531.